AI and Wind Power 2

AI and Wind Power 2

Advancing Sustainability, Grid Integration, and Future Frameworks

Edited by

Abhishek Kumar
Ananth Kumar T.
Ashutosh Kumar Dubey
Arun Lal Srivastav
J. Reyes Juárez-Ramírez

iSTE

WILEY

First published 2026 in Great Britain and the United States by ISTE Ltd and John Wiley & Sons, Inc.

ISTE Ltd
27-37 St George's Road
London SW19 4EU
UK

www.iste.co.uk

John Wiley & Sons, Inc.
111 River Street
Hoboken, NJ 07030
USA

www.wiley.com

Library of Congress Control Number: 2026935284

British Library Cataloguing-in-Publication Data
A CIP record for this book is available from the British Library
ISBN 978-1-83669-142-6

The manufacturer's authorized representative according to the EU General Product Safety Regulation is Wiley-VCH GmbH, Boschstr. 12, 69469 Weinheim, Germany, e-mail: Product_Safety@wiley.com.

Contents

Preface

The first volume of *AI and Wind Power* established a foundational truth: that artificial intelligence has become indispensable to the optimization of wind energy systems. From resource assessment and farm design to predictive maintenance and turbine control, we demonstrated how intelligent algorithms are transforming wind power from a passive generator of electricity into an active, adaptive and self-improving asset.

AI and Wind Power, Volume 2: Advancing Sustainability, Grid Integration and Future Frameworks was conceived to address this expanded horizon. This volume shifts the lens deliberately and decisively – from the component to the system, from the farm to the grid, from the present challenge to the future frontier. Our goal is to examine not only how AI makes wind power more efficient, but how wind power, augmented by AI, can transform the broader energy landscape. The chapters within this volume reflect the interdisciplinary nature of this undertaking. We begin with the smart grid itself, exploring AI-driven architectures for hybrid renewable energy systems that integrate wind with solar, hydro and storage. Contributors present sophisticated optimization frameworks – deep fuzzy-based particle swarm optimization, reinforcement learning-optimized deep neuro-fuzzy controllers – that dynamically balance supply and demand across diverse generation portfolios. These are not theoretical exercises, they are deployable methodologies tested against real-world variability and operational constraints.

Sustainability receives dedicated and rigorous treatment. Several chapters examine how AI can enhance sustainability across the wind energy lifecycle, extending beyond carbon accounting to encompass material efficiency, circular economy principles and the delicate trade-offs between energy output and

---

Preface written by Abhishek KUMAR, Ananth Kumar T., Ashutosh Kumar DUBEY, Arun Lal SRIVASTAV and J. Reyes JUÁREZ-RAMÍREZ.

environmental impact. Here, sustainability is framed not as a constraint on performance but as an optimization objective that AI is uniquely equipped to address.

The volume also ventures into territory that remains, for many, unexplored. Dedicated chapters examine the economic and market impacts of AI in wind power – how intelligent forecasting and automated trading are reshaping electricity markets, revenue models and investment decisions. Others explore the role of AI in policy and regulatory frameworks, asking not only how AI can comply with regulations but how it can inform their design. Still others investigate emerging frontiers: generative AI for hybrid system design, offshore wind in extreme environments and the application of AI to understand and influence energy consumption behavior.

Volume 2 deals with the intelligent system, whereas Volume 1 dealt with the intelligent machine. It is about how wind power is learning to communicate with solar arrays, battery storage, flexible demand and grid operators. It is about how sustainability is transitioning from an aspirational goal to an algorithmically tractable problem. It is about how economic value is created, captured and distributed in AI-augmented electricity markets. And it is about how we govern these technologies wisely – ensuring that the AI systems we deploy in service of the energy transition are themselves transparent, accountable and aligned with human development. We extend our sincere gratitude to each of the contributing authors for their intellectual rigor, their patience through multiple revisions and their willingness to venture beyond disciplinary comfort zones. We are also grateful to our co-editors for their complementary expertise and steady judgment, and to the editorial and production teams at ISTE Ltd for their unwavering partnership throughout this project.

April 2026

1

AI-Driven Advanced Smart Grid with Optimized Hybrid Renewable Energy Systems

1.1. Introduction

Hybrid renewable energy systems (HRES) combine multiple renewable energy sources, including solar, wind and energy storage, to provide a stable, clean and sustainable power source. By leveraging the complementary characteristics of various energy sources, HRES can minimize the reliance on traditional fossil fuels, enhance grid stability and maximize overall system reliability. Nonetheless, in spite of such advantages, HRES are fraught with inherent difficulties, such as the variability of renewable resources and energy demand and the intricacy of handling generation, storage and use simultaneously for the applications that we intend to use. Artificial Intelligence (AI) provides a revolutionary solution to overcome such difficulties by facilitating smart forecasting, control and optimization of HRES operations. Sophisticated methods, including machine learning (ML), deep learning (DL) and reinforcement learning (RL), enable these systems to forecast renewable generation and load demand accurately, adaptively regulate energy dispatch, and maximize storage use for the system under consideration, and are useful for many applications. AI-based optimization optimizes scheduling, cuts energy losses and lowers operating costs while enhancing the reliability and resilience of the power supply. The incorporation of AI into HRES has a significant effect, and not only

Chapter written by Mary A.G. EZHIL, S. JAISIVA, M. SUTHANTHIRA, R. ANUJA, M. Dhiviya NYCIL and A.S. MONIKANDAN.

is the technical performance improved but there are also compelling economic and environmental advantages. Through enabling accurate forecasting, component sizing at optimal levels and adaptive load management, AI-optimized systems can reduce the levelized cost of energy (LCOE) and minimize greenhouse gas emissions. With advanced smart grid infrastructures being developed, AI-optimized HRES are likely to play a central role in enabling the system practices by integrating the large-scale adoption of renewable energy, facilitating global decarbonization agendas and building a clean, self-sufficient and smart energy ecosystem for the future.

1.2. Overview of renewable energy systems

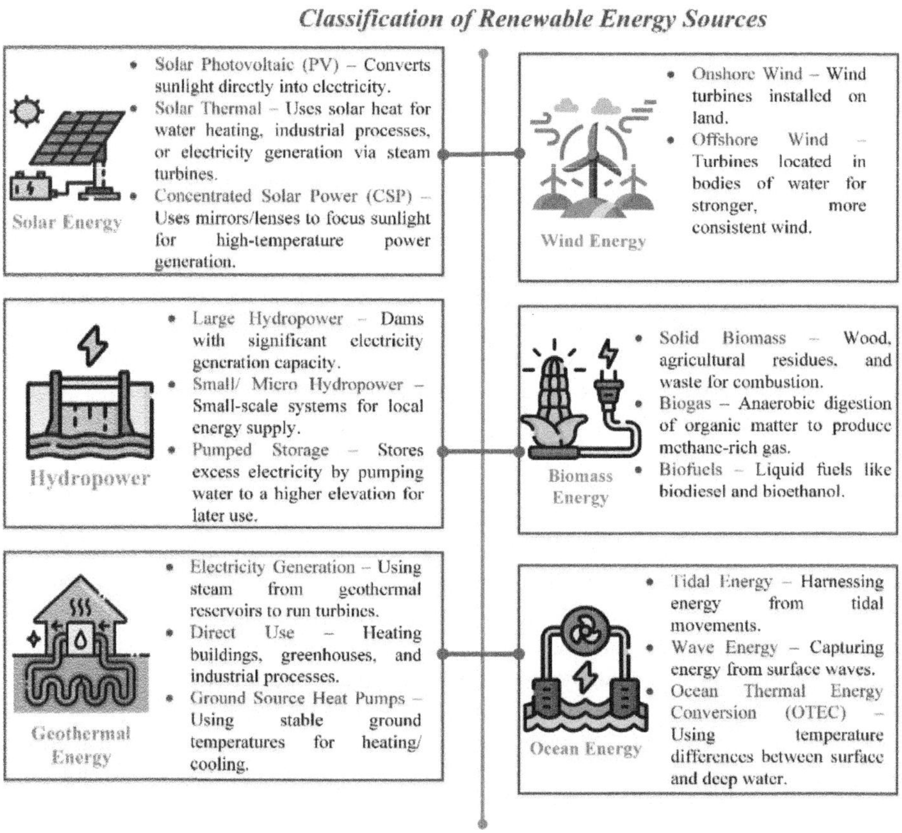

Figure 1.1. *Classification of renewable energy systems. For a color version of this figure, see www.iste.co.uk/kumar/windpower2.zip*

RES use natural, renewable resources such as sunlight, wind, water, geothermal heat and biomass to produce electricity and heat. In contrast to fossil fuels, these resources are readily available, sustainable and emit very low and nominal levels of greenhouse gas emissions, rendering them crucial in countering climate change and mitigating environmental degradation (Kothona et al. 2022c). RES technologies have developed substantially, considering all the forms, from large-scale solar farms and offshore wind farms to small-scale, decentralized systems for rural electrification. Their use is necessitated by the demands for cleaner energy alternatives, energy security and adherence to the international target for sustainability. Figure 1.1 shows the categorization of renewable energy systems.

The design and operation of RES involve integrating generation technologies with supporting infrastructure such as storage units, power electronics and grid interconnections. Modern RES increasingly leverage advanced forecasting, smart grid integration and hybrid configurations to overcome intermittency and improve reliability (Diagne et al. 2013).

As technological innovation, cost reductions and supportive policies converge, RES are becoming central to global energy transitions, enabling a shift towards cleaner, more resilient and economically viable energy networks for the future.

1.2.1. *Analysis of solar energy systems*

Solar energy systems use sunlight to generate electricity or heat, and they provide a non-polluting, renewable energy that is sufficient (Al-Hajj et al. 2018). In general, solar energy systems operate by capturing solar radiation using photovoltaic (PV) panels or solar thermal collectors, which then convert sunlight into usable energy for residential, commercial and industrial purposes. The classification and characteristics of solar energy systems are shown in Table 1.1.

Solar energy systems' scalability enables deployment from small rooftop systems to large-scale solar farms, making them a flexible way to meet energy needs, while helping to limit reliance on fossil fuels and reduce greenhouse gas emissions. The efficiency and performance of solar energy systems depend on site-specific factors such as location, solar irradiance, panel technology, tracking methods and storage (Voyant et al. 2017). By integrating solar energy with storage and other renewable energy sources to create hybrid systems, it is possible to more reliably stabilize energy output, thus positioning solar systems to further support sustainable and resilient energy infrastructure (Kothona et al. 2021).

Type	Description	Key advantages
Solar photovoltaic (PV)	It converts sunlight directly into electricity, using semiconductor-type materials.	Low maintenance, modular, scalable, silent operation.
Solar thermal	Uses solar energy to heat a fluid for direct heating or to produce steam for electricity generation.	Efficient for heating applications, it can be integrated with industrial processes.
Concentrated solar power system (CSP)	Using the mirrors/lenses to concave the focus area of sunlight onto a receiver system to generate high-temperature heat for power generation.	Can integrate with thermal storage for continuous generation; suitable for large-scale plants.

Table 1.1. *Classification and features of solar energy systems*

1.2.1.1. *Modeling of solar photovoltaic system*

The module's current–voltage (I–V) relationship is derived from the Shockley diode equation [1.1], with additional parameters incorporated in equation [1.2] to more accurately match the PV array's measured performance

$$I = I_0\left[\exp\left(\frac{qV}{kT}\right)-1\right] \quad [1.1]$$

$$I = I_L - I_0\left[\exp\left(\frac{q(V+R_s\cdot I)}{n\cdot k\cdot T}\right)-1\right]-\frac{V+R_s\cdot I}{R_{sh}} \quad [1.2]$$

where qV denotes the solar irradiance falling on the PV panel surface, while R_s represents the temperature-related effect on PV parameters (Abumohsen et al. 2024). R_{sh} is the voltage at maximum power, and JJ is the current at that point. Figure 1.4 shows the PV solar system model.

The variables considered include the ideality factor (n), series resistance (RS) and shunt resistance (RSh). These were determined using the manufacturer-provided short-circuit current, open-circuit voltage and maximum power point from the I–V characteristics. Figure 1.2 presents the I–V curves under standard test conditions and varying solar irradiance, highlighting the temperature effect (Liao et al. 2021).

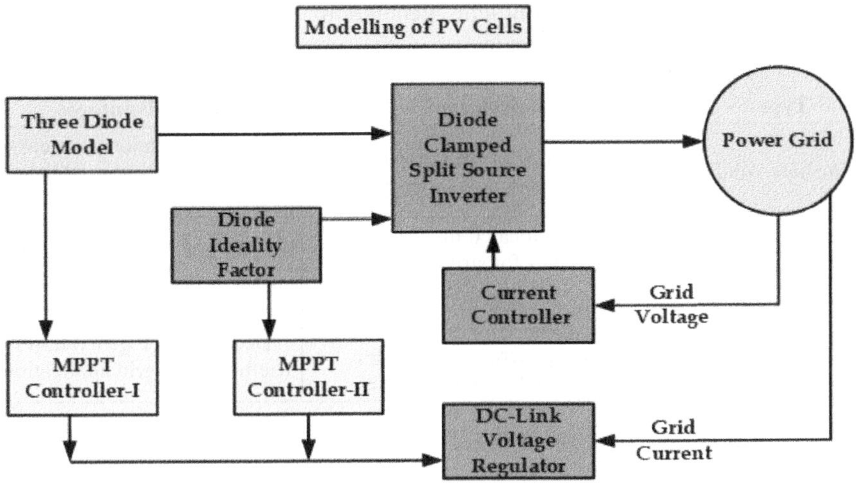

Figure 1.2. *Modeling of the PV solar system. For a color version of this figure, see www.iste.co.uk/kumar/windpower2.zip*

1.2.2. *Analysis of wind power systems*

Wind power systems represent an important step in diversifying the global energy mix and lowering reliance on fossil fuels. Both of these objectives will help mitigate climate change and greenhouse gas emissions. Wind power is a clean form of renewable energy that can be harnessed onshore and offshore, producing sustainable electricity without depleting our natural resources (Li et al. 2025). Wind energy is an increasingly viable alternative for meeting energy demands as world energy requirements continue to increase. Wind power development protects environmental resources and supplies a reliable energy option. Therefore, wind energy will be a sustainable energy alternative for many years to come. A wide share of electricity and energy security, and independence can be sustained when incorporating wind energy. Wind power systems are categorized and classified in Table 1.2.

Technological advances, policy mechanisms and declining generation costs are some of the components that are creating a growing demand for wind power. It has also become one of the fastest-growing renewable energy technologies. Present-day wind turbines, on average, are more efficient, have higher capacity factors and include intelligent controls for better integration with the electric grid. When used alongside storage or hybrid renewable solutions, wind power can also improve supply stability, balance intermittent generation and contribute to resilient, low-carbon energy transition strategies.

Type	Description	Key advantages
Onshore wind	Wind turbines are installed on land to harness wind energy.	Lower installation cost, easier maintenance and established technology.
Offshore wind	Turbines located in oceans or large lakes for stronger, more consistent wind.	Higher and more stable energy generation, minimal land use.
Small-scale/ micro-wind	Compact wind systems for localized power generation, often for residential or remote areas.	Decentralized generation is useful for off-grid applications.

Table 1.2. *Classification and features of wind power systems*

1.2.2.1. *Analysis of the wind power modeling*

We will begin by outlining the modeling of wind power generation. Wind turbines are designed to generate electricity over a wide range of wind speeds (Jiang et al. 2023). However, selecting an appropriate site remains the most crucial factor in their design and installation. The mathematical expression for the wind turbine, (W_t), is derived as follows:

$$W_t = \frac{1}{2}\beta\pi A^2 C_p(\varpi,\psi)\zeta_w^3 \qquad [1.3]$$

Here, ζ_w represents the wind speed in $(meter/\sec)$, C_p is the power coefficient, A denotes the rotor swept area, and β refers to the air density in (Kg/m^3). The power coefficient is defined as the ratio of the turbine's mechanical power to the total power available in the wind

$$\varpi = \frac{\delta_t A}{\zeta_w} \qquad [1.4]$$

Here, δ_t denotes the rotor's angular speed in $(rad/\sec)$, and (TT_t) represents the aerodynamic mechanical torque of the wind turbine, expressed as:

$$TT_t = \frac{1}{2}\frac{\beta\pi A^3 C_p(\varpi,\psi)\zeta_w^2}{\varpi} \qquad [1.5]$$

Finally, q_{res} is obtained from experimental measurements. Note that the turbine's internal characteristics are temperature-dependent; for example, parameters such as air density and generator resistance can change with temperature and should be accounted for in the model.

1.2.3. *Analysis of geothermal energy*

Geothermal energy uses the Earth's internal heat for electricity generation and direct heating. Geothermal energy is renewable, sustainable and reliable. The greenhouse gas emissions are typically much less than fossil fuel-based generation options (Saleem Afridi et al. 2021). Geothermal power plants typically have high capacity factors because they are not affected by weather conditions and are a reliable base-load option. However, geothermal power plants rely on specific locations because they need to be near a geothermal reservoir, which is typically near tectonic plates. The exploration phase can also become a costly aspect, and environmental impacts, such as induced seismicity and groundwater contamination, are concerns. The types and characteristics of the geothermal energy system are shown in Table 1.3.

Type	Description	Key advantages	Limitations
Dry steam plants	Steam from geothermal reservoirs is used directly to power turbines.	Simple design, high efficiency in suitable sites.	Requires high-temperature steam sources; limited geographic availability.
Flash steam plants	High-pressure hot water is rapidly depressurized to produce steam.	More common, they can handle high-temperature fluids.	Inefficient at lower temperatures; potential scaling/corrosion issues.
Binary cycle plants	Heat from the geothermal fluid is transferred to a secondary fluid with a lower boiling point.	Can use low- to moderate-temperature resources; wider applicability.	Lower efficiency compared to flash/dry steam plants.
Enhanced geothermal systems (EGS)	Artificial reservoirs are formed by fracturing hot dry rock and circulating fluids through it.	Expands geothermal potential to non-traditional sites.	Technically complex, high initial cost, risk of induced seismicity.

Table 1.3. *Types and features of geothermal energy systems*

New technologies, such as enhanced geothermal systems (EGS), binary cycle plants and improved drilling techniques, are further opening up the potential of geothermal power to more sites than just traditional high-temperature resource areas.

Geothermal technology can also be coupled with renewables to provide improved runtime stability while reducing reliance on intermittent resources. The capital cost of entering the geothermal space can be expensive, but the operational costs tend to be low, while the longevity of plants can be decades, making geothermal energy economically competitive in productive areas (Hu et al. 2024). Geothermal energy's dispatchable character and low carbon footprint suggest it is an important element of the global clean energy mix.

1.3. Evolution phases of AI in hybrid renewable energy systems

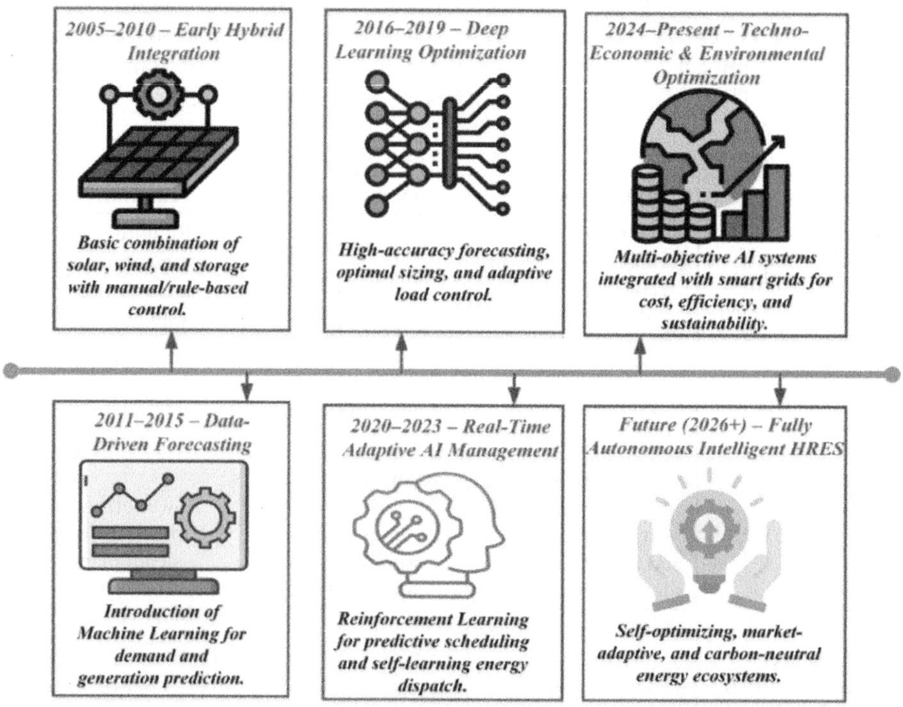

Figure 1.3. *Evolution phases of AI in hybrid renewable energy systems. For a color version of this figure, see www.iste.co.uk/kumar/windpower2.zip*

The development of AI in HRES marks a transition from simple hybrid configurations, where the human operator was often responsible for manually controlling energy elements, to fully autonomous, self-optimizing energy networks. Initially, the focus was on integrating and managing multiple renewable sources and mitigating intermittency as the first phase of development. This was followed in the second phase by implementing ML algorithms to improve forecasting, scheduling and advanced control of energy networks. The third phase further advanced this research through the use of DL algorithms for optimization, RL for real-time adaptive management and multi-objective optimization algorithms for techno-economics or environmental advantages. The current phase proposes the design and implementation of intelligent, market-driven energy networks that are fully embedded in smart grids to maximize the benefits to society and sustainability.

Figure 1.3 shows the evolution of AI integration in HRES, showing an incremental path from hybrid energy configurations to fully autonomous intelligent energy ecosystems (Al-Ali et al. 2023). The first phase is the basic integration or hybridization of renewables such as solar and wind, along with classical storage systems, with control being manual or rule-based. Table 1.4 shows the timeline of developments of AI-based hybrid RES.

Period	Advancements/key developments
2005–2010	Initial hybrid integration of solar, wind and basic battery storage. Manual or rule-based energy management. Focus on feasibility studies and operational proof-of-concept.
2011–2015	Introduction of ML techniques for generation and demand forecasting. Automated day-ahead scheduling to minimize the mismatch between supply and demand.
2016–2019	Adoption of DL and LSTM models for high-accuracy forecasting. AI-driven optimal component sizing (solar capacity, wind turbines, batteries) and adaptive load control for improved grid stability.
2020–2023	Deployment of RL for real-time adaptive energy management. Predictive scheduling for charging/discharging cycles, automated load shedding and fault detection systems.
2024–present	Multi-objective AI optimization balancing cost, efficiency and environmental impact. Seamless integration into smart grids for dynamic energy trading and improved reliability.
Future (2026+)	Fully autonomous, self-optimizing HRES capable of real-time adaptation to weather, market prices and grid demands. Achieves maximum renewable penetration, minimal downtime and carbon-neutral operation.

Table 1.4. *Periodic advancements in AI-driven hybrid RES*

The second stage involves the incorporation of data-driven methods, as ML models will be used in the generation and demand forecasting, in turn, enhancing scheduling and decreasing the energy mismatch. In the third stage, DL techniques will further develop in the form of neural networks and LSTM models as instruments for finding better forecasting, by optimizing component sizing and being able to conduct load control to improve overarching grid stability. In the fourth stage, there will be RL in real-time adaptive management approaches for scheduling prediction, automation and fault detection, similar to non-renewable energy generation models, increasing operational accuracy by means of optimization and educational models. In the fifth stage, we note the retention of multi-objective optimization theory, managing techno-economic performance, environmental sustainability and reliability simultaneously, as well as expanding implicitly to be navigated with the smart grid. In the final stage, this research will eventually lead to a fully autonomous HRES, that is fully self-optimizing and able to respond to weather patterns, market signals and grid requirements, approaching maximum penetration of renewables, at nearly-zero downtime.

1.4. Integration of AI in hybrid energy systems

The implementation of AI in hybrid energy systems (HESs) showcases a breakthrough towards an effective, dependable and sustainable method for good energy management. Hybrid systems that integrate various energy sources, such as solar, wind and storage, maximize generation potential, reliability and continuity of supply. Despite hybrid systems leveraging multiple energy sources, they are also susceptible to unique challenges regarding the variability of renewable resources, demand variability and the coordination of multiple and diverse resources. AI addresses these challenges through sophisticated forecasting, predictive analysis and adaptive control (Kothona et al. 2020). In addition, AI systems are capable of using historical data and real-time data to make accurate forecasts of renewable generation and demand, dispatch the available resources or decide what resources and storage to manage to balance supply and demand. AI delivers capabilities and functionality above basic controls so systems can perform things such as dynamically manage load, detect faults and conduct techno-economic optimization to ensure the hybrid system is at the highest level of performance and operating most cost-effectively concerning the environment. For example, ML provides the capability for data-driven decisions, DL provides the capability for improved forecasting and RL provides the capability for real-time adaptive control systems that will learn and improve system performance with time.

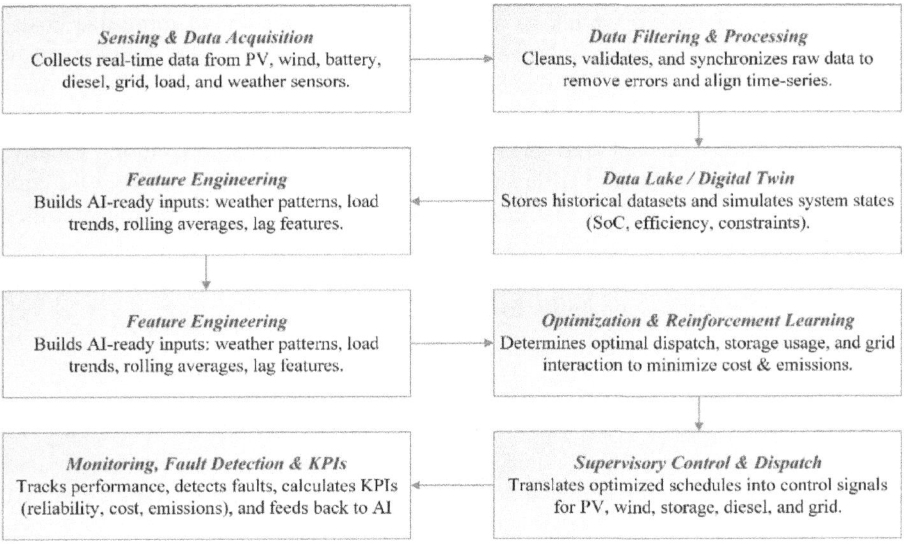

Figure 1.4. *Block diagram of AI-enabled hybrid energy system architecture*

Figure 1.4 depicts the block diagram of an AI-enabled HES architecture, which shows the flow of functions describing how AI integrates and enables hybrid energy sources and hybrid energy storage systems to optimize energy management. The whole process begins with sensing and data acquisition. In this first step, real-time data on many different entities, such as energy production from solar PV and wind, energy storage in batteries, diesel, utility grid, energy consumption for the load, as well as weather such as irradiation and wind speed, are captured continuously as raw data. This raw data is filtered and processed, which identifies bad or missing signals while also synchronizing the time-series data. The processed data streams are sent to a data lake as well as a digital twin of the HES, which can contain various aspects of the system including historical data and statistically relevant historical data, as well as current information of states, such as state of charge, efficiencies of the equipment, etc. Finally, the feature engineering block produces AI-ready data streams as derived data, which includes energy load patterns, weather features, or load and weather statistical indicators. These AI-ready data streams are then forwarded to a forecasting engine, which implement ML/DL models to forecast the following: renewable generation, demand and electricity price. The forecasts are used by the optimization and RL sub-module, which generate or determine the optimal dispatch and scheduling strategies, which, in short, lead the AI engine to minimize costs, maximize reliability and minimize emissions through battery and energy management, scheduling and dispatch. The supervisory control and dispatch unit implements the selected setpoints by issuing commands to physical assets, including solar, wind, battery, diesel and grid interface, for real-time operation. Finally, monitoring, fault detection and KPIs ensure that the system operates reliably through the identification of anomalies, estimates techno-economic indicators and provides information to refine forecasting and optimization models. Combined, these eight functional blocks form the AI-driven framework that ensures reliability, cost-effectiveness and sustainability for HESs.

1.4.1. *Classification of AI in renewable energy systems*

AI's integration in RES is altering how power generation, distribution and storage interact with consumers and electrical energy. In the vast range of AI techniques, neural networks are distinct because they learn complex patterns, accommodate nonlinear relationships and respond to operating conditions that can vary. Below, we present five categories of neural networks commonly applied in RES, each of which has its strengths for specific operational scenarios.

1.4.1.1. *Feedforward neural networks (FNNs)*

FNNs are often used in renewable energy systems (RES) for applications such as predicting solar irradiance, load forecasting and energy demand estimates. For example, an FNN could be trained on historical weather data and system output to predict daily solar photovoltaic generation (Manbachi et al. 2016). FNNs can be beneficial because of their ease of use, training time and deployment speed for timely decision-making in hybrid systems. They might struggle with time-series dependencies if not supplemented by other models.

1.4.1.2. *Convolutional neural networks (CNNs)*

CNNs were designed initially for image processing and are particularly good at extracting spatial features from data, which means that there are many potential applications of CNNs in renewable energy. Specifically, CNNs can be used to process satellite images or drone footage to monitor solar farms and identify dust or shading on PV panels and mechanical wear on wind turbine blades. CNNs could also be used to analyze geographical and meteorological maps for resource allocation for HESs. One of the advantages of CNNs, when compared to traditional ML methods, is that CNNs do not require preprocessing by expert human technicians to identify the appropriate features automatically. Consequently, CNNs can be extremely useful for fault detection as well as predictive maintenance to keep downtime to a minimum and boost energy yield.

1.4.1.3. *Recurrent neural networks (RNNs)*

RNNs are particularly useful for sequential or time-related data, as the output depends on all the previous inputs in addition to the current input. In renewable energy, RNNs are useful for applications such as forecasting wind speed in short time frames, forecasting energy prices and managing dynamic load balancing.

An example of this would be an RNN processing the past wind speed measurements to create a forecast of the coming wind speed to be used to schedule hybrid systems accordingly. However, RNNs can potentially fall victim to the vanishing gradient problem, limiting their goal for long-term dependencies. Breaking the vanishing gradient limit is possible with a different neural network framework called a long short-term memory (LSTM).

1.4.1.4. *Long short-term memory (LSTM) networks*

An LSTM integrates memory cells into an RNN, enabling it to exploit long-term memory to recognize patterns across multiple time series. LSTMs are very useful for seasonal forecasting as well as long-range forecasting, for example, forecasting solar energy production over several months or anticipating wind variability associated with monsoons. LSTMs adapt well to hybrid RES as they can, in advance, anticipate and plan how to use battery storage and predict excess and deficit of energy optimally. LSTMs also have the advantage of using long chunks of data to understand the longer-term picture, allowing researchers to use them for strategic energy planning and resource optimization.

1.4.1.5. *Radial basis function networks (RBFNs)*

RBFNs differ from other neural networks because they use radial basis functions (typically Gaussian functions) as an activation function. They are suitable for non-trivial regression, function approximation and pattern classification. In the renewable energy sector, RBFNs are applicable to optimize microgrid operation, power flow control and fault detection in complex hybrid systems. The benefit of using RBFNs is based on fast training times and strong generalization abilities, which can be leveraged for real-time control applications in energy management systems.

Table 1.5 depicts the comparative analysis of neural network architectures and their applications in renewable energy systems.

NN Type	Core feature	Applications in RES	Advantages	Limitations	Numerical insights (general)
Feedforward Neural Networks (FNN)	Basic forward connections, no memory of past inputs.	Solar irradiance prediction, load forecasting, energy demand estimation.	Simple structure, fast training, quick deployment, good for short-term decisions.	Struggles with sequential/time-series data, limited temporal learning.	- Forecasting accuracy: ~85–90% (short-term solar). - Training time: Few seconds to minutes for small datasets (~10k samples). - Deployment latency: < 50 ms per prediction.
Convolutional Neural Networks (CNN)	Extracts spatial features using convolutional filters.	Image-based RES monitoring (e.g. dust/shading detection on PV panels, wind turbine blade wear, geographical resource allocation).	Automatic feature extraction, high accuracy for spatial data, strong in-fault detection and predictive maintenance.	Computationally expensive; requires large datasets; less suited for sequential data alone.	- Fault detection accuracy: ~90–95% (panel/wind blade images). - Training dataset size: Typically > 50k labeled images .- Inference time: ~100–200 ms per image (with GPU).
Recurrent Neural Networks (RNN)	Sequential learning, output depends on current + previous inputs.	Short-term wind speed forecasting, real-time energy price prediction, dynamic load balancing.	Good at handling sequential data, adaptive to changing inputs.	Vanishing gradient problem for long-term dependencies, limited memory span.	- Forecasting accuracy: ~80–88% (short-term wind speed). - Effective prediction horizon: 0–24 hours. - Training epochs required: ~100–300 for convergence.

Long Short-Term Memory (LSTM)	RNN variant with memory cells for long-term dependencies.	Seasonal/long-range forecasting (solar over months, wind during monsoons), storage optimization, strategic energy planning.	Handles long-term dependencies, robust to vanishing gradient, useful for strategic forecasting.	Slower training, higher computation cost compared to FNN/RNN.	- Forecasting accuracy: ~90–95% (seasonal solar/wind). - Prediction horizon: Days to months. - Training time: Minutes to hours (depends on dataset size, ~100k samples).
Radial Basis Function Networks (RBFN)	Uses radial basis (Gaussian) activation; fast function approximation.	Microgrid optimization, power flow control, real-time fault detection.	Fast training, strong generalization, efficient for real-time energy management.	Less effective for very large datasets, requires proper center selection.	- Control accuracy: ~88–92% (grid operation). - Training time: < 1 min for moderate dataset (~20k samples). - Suitable for < 10 ms real-time response in energy control.

Table 1.5. *Comparative analysis of neural network architectures and their applications in renewable energy systems (RES)*

1.5. Analyzing the integration of AI models in renewable energy systems

In the context of an AI-optimized HRES, clustering entails grouping distributed energy generation units and storage elements into operational units that operate in coordination with one another and communicate performance and status information to the cluster controller (CC), which assesses the status of each cluster and communicates to CEMS in a hierarchical structure. The goal of clustering is to organize assets in the system to form operational groups that take into account underlying patterns of energy production, available resources and demand profiles. The operational clusters in this study were identified using a dynamic fusion clustering (DFC) algorithm.

Deep fuzzy clustering

In the proposed AI-HRES framework, fuzzy clustering and the deep neural network's (DNN) representation learning capabilities are leveraged by the DFC algorithm to achieve efficient grouping of distributed energy resources, such as solar PV arrays, wind turbines and battery storage units (Karagiannopoulos et al. 2019). The fuzzy membership in DFC uses two cluster matrices: the fuzzy within-cluster scatter matrix ς_{wf}, representing operational similarity among assets with comparable solar irradiance profiles or energy production characteristics, and the fuzzy between-cluster scatter matrix ς_{lw}, capturing the operational distinctiveness between clusters under varying renewable resource availability and load demand scenarios. These matrices form the mathematical foundation for optimizing energy flow coordination and maximizing renewable penetration in the system. The cluster matrices can be expressed as:

$$\varsigma_{lf} = \sum_{t=1}^{c} \sum_{j_1=1}^{T} p'^{m_1}_{tj_1} (a_t - B_{j_1})(a_t - B_{j_1})^{\beta} \quad [1.6]$$

$$\varsigma_{lw} = \sum_{t=1}^{c} \sum_{j_1=1}^{T} i'_{j_1} p'^{m_1}_{tj_1} (a_t - \bar{B})(a_t - \bar{B})^{\beta} \quad [1.7]$$

wherein the amount of the AI system is denoted as T, the fuzzifier is indicated as m_1, a_t represents the tth data, B_{j_1} specifies the j_1^{th} center, $\bar{B} = 1/c \sum_{t=1}^{c} a_t$

indicates the average of every information system, the fuzzy solar region function is designated by $p'_{\iota j_1}$, the batch is denoted by c and i'_{j_1} indicates the parameter

$$Loss(p'_{\iota j_1}, B_{j_1}) = \sum_{\iota=1}^{c}\sum_{j_1=1}^{T} p'^{m_1}_{\iota j_1} \| a_t - B_{j_1} \|^2 - \sum_{\iota=1}^{c} i'_{j_1} \sum_{j_1=1}^{T} p'^{m_1}_{\iota j_1} \| a_t - B_{j_1} \|^2 \quad [1.8]$$

For creating the term $p'_{\iota j_1}$ differentiable to $Loss(p'_{\iota j_1}, B_{j_1})$, the parameter i'_{j_1} is signified beneath the RES and with the corresponding AI system to be considered,

$$i'_{j_1} = \frac{(\vartheta/4)\min_{j_1' \neq j_1} \| B_{j_1} - B_{j_1'} \|^2}{\max_r \| j_{1r} - \bar{j}_1 \|^2} \quad [1.9]$$

wherein ϑ denotes the renewable energy system hyper-parameter that is used to balance the distance within and between notified clusters (solar/wind). The fuzzy membership function $p'_{\iota j_1}$ can be expressed as follows:

$$p'_{\iota j_1} = \frac{\left(\| a_t - B_{j_1} \|^2 - i'_{j_1} \| B_{j_1} - \bar{B} \|^2\right)^{-1/(m_1-1)}}{\sum_{r=1}^{T}\left(\| a_t - B_r \|^2 - i'_r \| B_r - \bar{B} \|^2\right)^{-1/(m_1-1)}} \quad [1.10]$$

As a result, the different nodes are successfully grouped by the AI-induced technique under an overall of T sets, and the cluster that is thus created is designated by C'.

1.5.1. *Analyzing the optimization algorithms in the integration of AI into renewable energy systems*

To successfully integrate AI into RES, advanced optimization algorithms will be required to deal with the complexity, improve performance and optimize the efficiency of the whole system. Optimization algorithms provide the necessary ability to optimize the limits of variable inputs, use resources efficiently, reduce operational costs and ensure systems remain stable when integrated with hybrid renewable energy sources. A good optimization process will be critical for decisions affecting major aspects of RES design and use such as sizing components, predicting power availability and load forecasts, and managing energy storage. It is these algorithms that ensure automated AI energy systems remain operational with little human management while being adaptive to changing environmental and demand conditions. There are many kinds of optimization algorithms that can generally be categorized by the way they work, the degree of flexibility they have and how they approach problem solving.

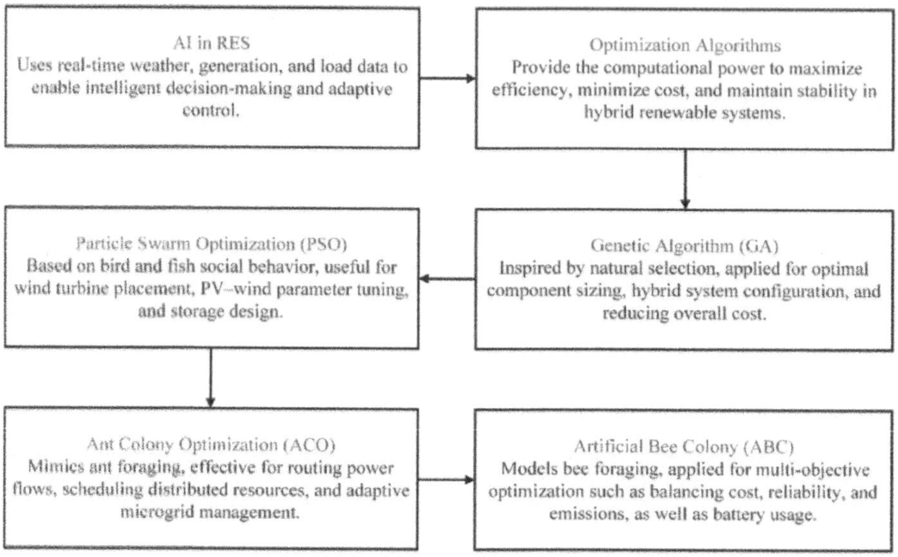

Figure 1.5. *Block diagram of AI-driven optimization algorithms for renewable energy systems*

Figure 1.5 shows a block diagram of AI-led optimization algorithms for renewable energy systems, and highlights how AI is incorporated into optimization strategies for better performance. The first aspect to consider is AI in RES, where data is transformed into intelligent decisions using real-time actual weather conditions, load demand and renewable generation. The data is routed through optimization algorithms, seen here as the processing "brain", to improve performance through efficiency, cost savings, reduced risk, stability, etc. We look at five evolutionary algorithms that play a fundamental role in being able to do this: a genetic algorithm (GA) uses the principles of natural selection, genetic combination, mutation and fitness of a solution to size components and minimize costs; particle swarm optimization (PSO) takes into account patterns of social behavior, optimizes wind turbine layouts, performs parameter tuning and conducts storage designs; ant colony optimization (ACO) is a multi-agent metaheuristic algorithm that mimics ant foraging behaviors to perform routing and schedule optimization of microgrids; artificial bee colony (ABC) imitates the food foraging behaviors of honey bee swarms to provide a trade-off management process of cost, reliability and emissions; and differential evolution (DE) uses mutation and crossover techniques to tweak forecasts, optimize dispatch and performance. These algorithms ultimately join together to provide an optimized operation of a renewable energy system, resulting in reliable, responsive and cost-effective decision-making, with minimized human intervention.

A classification of these five types that are especially applicable to renewables using AI is given below.

1.5.1.1. *Genetic algorithms (GAs)*

GAs, inspired by the mechanics of natural selection, represent evolutionary optimization methods. GAs work by first generating a number of potential solutions (a population), evaluating their performance (the fitness of each) and applying genetic operators such as selection, crossover and mutation to evolve better solutions over generations. In relation to RES, GAs are most frequently used for optimal sizing of components, configuration of hybrid systems, as well as load management (using batteries) schemes to minimize costs or maximize efficiency.

1.5.1.2. *Particle swarm optimization (PSO)*

PSO is a population-based algorithm inspired by the social behavior seen in birds and fish. In PSO, each "particle" represents a potential solution, and they move through the search space according to their own experience and that of their neighbors. PSO is effective for continuous optimization problems, for example, predicting optimal wind turbine placement, optimizing parameters in PV arrays and optimizing hybrid storage system design for increased reliability and lower costs.

1.5.1.3. *Ant colony optimization (ACO)*

ACO is modeled after the foraging behavior of ant colonies, where ants use pheromone trails to navigate back to their nests from sources of food. ACO is useful in RES for path optimization in energy distribution networks, scheduling of distributed generation resources and adaptive routing in microgrids. Ants can adapt dynamically to changing conditions, so it is advantageous for applications in real time or smart grid energy management.

1.5.1.4. *Artificial bee colony (ABC) algorithm*

ABC algorithms imitate the foraging behavior of honeybees and contain employed bees, onlooker bees and scout bees that browse and exploit search spaces. The ABC algorithms are applied in hybrid RES for multi-objective optimization problems to manage the conflict between cost, reliability and emissions. They can be particularly effective in parameter tuning for controllers and in optimizing battery management systems.

Algorithm	Inspiration	Working principle	General strengths	RES/wind power applications
Genetic algorithm (GA)	Natural selection and genetics	Generates population of solutions, uses selection, crossover, mutation to evolve better ones	Good for discrete/nonlinear problems, handles multi-objective tasks	Optimal sizing of wind turbines and hybrid RES, load management with batteries, minimizing cost of wind–solar–storage systems
Particle swarm optimization (PSO)	Social behavior of birds/fish	Particles adjust positions based on personal and neighbors' best experiences	Fast convergence, simple implementation, effective in continuous search spaces	Wind turbine placement optimization, tuning parameters of PV/wind models, hybrid storage system reliability improvement
Ant colony optimization (ACO)	Ant foraging with pheromone trails	Uses artificial "ants" that explore and reinforce best paths	Dynamic adaptation, effective in routing and scheduling	Optimal routing in wind-integrated microgrids, distributed generation scheduling, real-time energy distribution with wind power
Artificial bee colony (ABC)	Foraging behavior of honeybees	Used, onlooker, and scout bees explore/exploit solutions	Balances exploration & exploitation, good for multi-objective trade-offs	Optimizing wind–battery operation, controller parameter tuning, reducing cost versus reliability conflicts in hybrid RES
Differential evolution (DE)	Evolutionary mutation and crossover	Creates new candidate solutions via scaled differences of existing ones	Robust in continuous, nonlinear, multimodal problems	Forecast model parameter optimization for wind, dispatch strategy optimization in wind–storage, hybrid system configuration for long-term performance

Table 1.6. *Comparative overview of evolutionary optimization algorithms in RES*

1.5.1.5. *Differential evolution (DE)*

DE is a strong, yet simple algorithm that uses mutation, crossover and selection to complete optimization tasks. DE can be widely used for continuous, nonlinear and multimodal optimization problems. In the context of renewable energy, DE has been used to optimize parameters in predictive models, optimize dispatch strategies for energy storage or determine perfect design configurations for HESs so long-term performance can be strengthened. Table 1.6 depicts the comparative overview of evolutionary optimization algorithms in RES.

1.6. AI-optimized hybrid system design

All of these phases encapsulate the entire process of deploying AI in hybrid RES, starting from identifying relevant information sources and preprocessing data to ensure legitimate, clean and necessary input. Next is the phase of selecting and training a model, where the essential goal is to choose appropriate AI frameworks/architectures and learning strategies to propose a solution to minimize the system. Next is the phase of ongoing optimization parameter tuning, a cyclical process of attempting to optimize the proposed model's performance based on attaining optimality under the constraints of accuracy, efficiency and adaptability. Last is a phase of simulation, validation and implementation of the prototype in real time in order to operate reliably in a practical setting with all associated assumptions and limitations while achieving its energy, cost and sustainability goals. The workflow for AI-enhanced hybrid RES is shown in Figure 1.6.

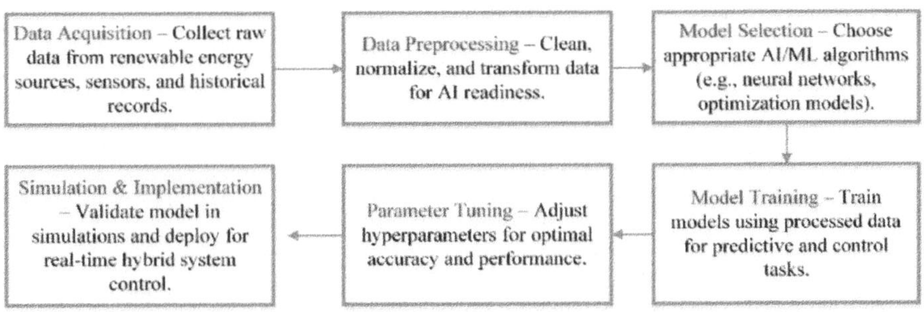

Figure 1.6. *Workflow for AI-optimized hybrid RES*

1.6.1. *Data acquisition and preprocessing*

Data acquisition and data preprocessing are the basis of any AI-based optimization framework for hybrid RES. The quality and accuracy of any AI models implemented in the context of renewable energy applications depend on the diversity, accuracy and completeness of the input data. The term data acquisition refers to the structured collection of relevant data and information from multiple sources. Due to the unstructured nature of acquisition, preprocessing is required to clean, structure and transform the data into a format that can be used in subsequent analysis and model training. In an HRES context, data acquisition will use multiple sources including meteorological sensors (i.e. solar irradiance meters, anemometers and temperature sensors), smart meters, SCADA systems, GIS information and publicly available datasets (i.e. NASA POWER, NREL's NSRDB and European Copernicus services). In these situations, a range of relevant acquired parameters includes solar radiation intensity, wind speed, ambient temperature and humidity, geothermal gradient, load demand specifications and grid specifications. Historical records of data are provided with real-time streaming data. In these instances, the development of predictive and adaptive control strategies requires historical and real-time data. Table 1.7 depicts the key steps in data preprocessing for AI model development.

Once data is collected, preprocessing becomes essential to ensure data integrity and usability. The preprocessing stage includes several key steps.

Step	Description	Example/techniques
1) Data cleaning	Removal of erroneous readings, missing values, outliers and inconsistencies caused by sensor malfunctions or communication errors.	Interpolation, statistical imputation, anomaly detection
2) Data transformation	Conversion of raw measurements into a uniform scale and format.	Converting wind speed to m/s, normalizing temperature data for model input
3) Feature engineering	Derivation of additional relevant parameters (features) from the raw data.	Calculating solar zenith angle from timestamp and location, computing moving averages of wind speed
4) Data normalization and scaling	Ensuring all input features have comparable ranges to prevent bias in AI algorithms.	Min–max scaling, Z-score standardization
5) Data Segmentation	Organizing data into training, validation and testing subsets for AI model development, evaluation and deployment.	Splitting the dataset into 70% training, 15% validation and 15% testing

Table 1.7. *Key steps in data preprocessing for AI model development*

Effective preprocessing not only enhances the predictive capabilities of AI models but also maximizes stability in real-time decision-making for HRES operation. For hybrid systems, effective preprocessing is crucial because of the variability and complexity of integrating multiple types of renewable energy sources. With sufficient data acquisition and effective preprocessing, AI-based HRES will achieve a higher accuracy in forecasting, drain minimal resources and have the best energy allocation management.

1.6.2. *Model selection and training*

Choosing the right AI model is an important first step in developing an efficient HRES. Depending on the specific objectives of the system to be designed – be it to forecast renewable generation, optimize dispatch strategies, forecast maintenance requirements or manage multiple energy sources – the model must be appropriate to those objectives. Artificial neural networks (ANNs), spatiotemporal artificial neural networks (CNNs), long short-term memory (LSTM) networks for time-series prediction and hybrid models that combine ML with metaheuristic optimization methods are common AI models used in HRES cases. The model selection process begins by clarifying the problem type, whether it be classification, regression or clustering, and also the availability, quality and dimensionality of the data.

An example of this is that forecasting solar output requires a DL approach that has a temporal aspect, while optimizing power flows may require RL or evolutionary computation methods. Once the nature of the model has been decided, its training involves feeding the model with processed data and repeatedly adjusting the model's parameters, via iterative techniques, to decrease prediction or optimization error. Every model has iterative techniques, such as backpropagation (for neural networks), gradient-based optimizers (e.g. Adam, Knuth transform) and cross-validation. Regularizations (dropout, L2 penalties) will be applied to restrict over-learning and ensure that the model generalizes well to unseen data.

1.6.3. *Optimization parameter tuning*

Parameter tuning is also referred to as hyperparameter optimization. It is the process of adjusting the control variables of an AI model, or optimization algorithm so as to obtain the peak performance. Indeed, for AI-optimized HRES, the performance of forecasting, scheduling or dispatch algorithms will depend on the proper choice of hyperparameters, including the learning rate, number of layers in a neural network, mutation rates in genetic algorithms and inertia weights in PSO.

Methods for performing parameter tuning are:

1) Grid search – testing every combination in a specified parameter space.

2) Random search – sampling parameter values at random from defined ranges, faster than grid search, particularly for high-dimensional problems.

3) Bayesian optimization – leveraging probabilistic models to make a prediction about the best parameters you can use for your AI-optimized HRES, based on previous evaluations.

4) Evolutionary strategies – using meta-heuristic optimization methods to fix optimization parameters.

In HRES, it is essential to tune parameters in order for AI models to represent nonlinearities between renewable generation, demand and storage variations. A poorly tuned value can lead to bad forecasting accuracy, energy dispatch not being optimal or calculations that are inadequate for the intended purpose. Consequently, especially in real-time applications, automated parameter tuning methods, with feedback loops for performance, are often selected over other tuning techniques.

1.6.4. *Simulation and validation*

Before implementation of an AI-based optimization model in a live HRES environment, simulation and validation are very important. Simulation attempts to reproduce the conditions of the hybrid system's operation on virtual testbeds, allowing AI algorithms to be trialed with varying environmental conditions and loads in a safe space without endangering operational stability. Simulation tools such as MATLAB/Simulink, HOMER Pro and DIgSILENT PowerFactory may allow for the use of AI modules to see how a model may operate in renewable generation forecasts, storage control functions and grid interaction. These tools can simulate deterministic and stochastic simulations to account for the uncertainty of renewable energy availability, load demand and other conditions.

Validation ensures that the AI model's outputs align with real-world measurements and operational constraints. This process often involves:

– cross-validation using historical datasets;

– out-of-sample testing to evaluate generalization;

– sensitivity analysis to determine the robustness of the model to parameter changes;

– performance metrics such as mean absolute error (MAE), root-mean-square error (RMSE) and system reliability indices.

By validating models under various stress scenarios such as sudden weather changes or component failures, researchers can ensure that the system remains reliable, resilient and efficient.

1.6.5. *Real-time implementation considerations*

Using AI-optimized models for HRES in real-time settings has practical challenges. Real-time operation involves the acquisition of data of very low latency, performing analysis in rapid timeframes and adapting to constantly changing operational conditions. It brings with it the need for solid communication infrastructure, edge computing capabilities and redundant hardware to avoid outages. The important practical considerations include:

1) Integration with SCADA systems – making sure data flows seamlessly between sensors and controllers and the AI algorithms for decision-making.

2) Computational maintenance – using either lightweight AI models or edge AI to run all as close to the data as possible, mitigating latency in submission of data to the server for processing.

3) Fault tolerance and redundancy – failover mechanisms can help maintain some form of operation, even when hardware or networks fail.

4) Cybersecurity – AI-optimized HRES will require devices communicating with data streams, so protecting the communications and control systems must occur. Normal standards for securing communications involving encryption, secure authentication and common cyber threats through anomaly detection will be essential.

5) Regulatory concerns – AI control frameworks will need to comply with codes and regulations for their energy grid, environmental regulation and safety.

In real-time HRES applications, AI models need to be self-updating, leveraging the latest data, to account for unexpected changing conditions such as changes in clear sky solar irradiance or wind speed. It is important to remember that successful implementation is dependent on proper operation and capability with the algorithms and their overall accuracy, as well as stability within the larger system of operation, proper operator training and support, and the mechanisms to monitor it robustly.

1.7. Performance metrics and evaluation

Performance metrics and evaluation offer a formalized means for assessing the performance of AI-enhanced hybrid RES with respect to effectiveness, efficiency and reliability. Performance metrics provide a basis for stakeholders to quantify energy output, resource productivity, cost savings and operational stability and help create an evaluation framework that embraces AI or ML analytics. Research proposes that the evaluation transforms into an interactive process, consisting of real-time performance monitoring, predictive performance information and an open loop for iterative performance adjustments within the performance metrics of various research disciplines. For a comprehensive evaluation framework to be proposed, researchers realize that there are different forms of evaluation: technical indicators, economic indicators and resilience-based indicators, where technical indicators are related to real-time energy performance. It is equally important to recognize that the utility of technical indicators is uneven when faced with the possibility that future performance goals may alter to account for environmental change or demand volatility.

1.7.1. *Energy efficiency and loss reduction*

Energy efficiency in AI-optimized hybrid RESs is simply how well the system converts the work done by renewable resources into usable power and limits all energy waste. AI algorithms improve efficiency by optimizing loads for dispatch over alternative storage options, limiting idle periods for generating equipment and ultimately reducing total energy loss from transmission and conversion. Continuous performance monitoring with adaptive control strategies can ensure energy efficiency levels reach optimal points based on varying environmental and demand situations at any given time. The direct efficiency trust of an energy asset is used to determine the efficiency of electricity delivery by evaluating the asset capacity to transfer as much power efficiently with the least contribution to energy loss in any given context. Direct efficiency trust in the context of energy assets is the probability that power from inseparably interconnected assets/functions transfers successfully to the defined load or storage entity under ideal operating conditions. In an AI-optimized control architecture, this trust represents the combined factors of technical performance (i.e. low line losses, superior power quality) and the reliability of the asset subsequently established through predictive analytics. Direct efficiency trust can be calculated in the following way:

$$D_{trust}(W_\eta, W_z) = \frac{b_{\eta z}(p)}{b_{\eta z}(p) + C[\mu_{\eta z}(p) - b_{\eta z}(p)]} \quad [1.11]$$

Here, $b_{\eta z}(p)$ indicates the total amount of energy (in kWh) successfully transmitted by the energy unit W_η, $\mu_{\eta z}(p)$ represents the amount of energy delivered from the distribution point W_η to the load node W_z and p denotes the evaluation period of the unit's performance. The penalty weight associated with all inefficient or loss-inducing components is represented by C, and the direct efficiency trust is specified as D_{trust}.

1.7.2. *Cost–benefit analysis*

Cost–benefit analysis provides an evaluation of the economy of deploying AI-integrated hybrid systems, initially by comparing required capital and operational expenditures against anticipated long-term value-like financial returns. Benefits may include fuel savings, reduced maintenance costs, extended asset lifespan and increased penetration of renewables. AI-enabled predictive analytics also assist in the position, identification and forecasting of energy demand and market values and inform AI-optimized energy consumption decision-making. This represents the confidence in a system component's ability to assess the cost–benefit and reliability of some other component before making a recommendation to allocate particular resources or prioritize operations. Components that do not lend themselves to frequent transactional exchanges rely on cost–benefit information provided through the component adjacency of other components, in order to allocate resources, which could be derived as,

$$ID_{trust}(p) = \sum_{\gamma=1}^{h} n_\gamma * U^{\gamma}_{\eta'z'}, \ \gamma = 1, 2, ..., h \qquad [1.12]$$

Here, n_γ specifies the weighting factor of $U^{\gamma}_{\eta'z'}$, which represents the evaluated cost–benefit trust value from asset η' to asset z', U indicates the overall trust score in terms of operational cost-effectiveness, h specifies the number of selected assets or operational units considered in the analysis, and the indirect cost–benefit trust is specified as ID_{trust}.

1.7.3. *Reliability and resilience indicators*

Reliability indicators evaluate the system's ability to provide energy outcomes consistently, without interruptions, while resilience indicators quantify the capacity to remain functional (and even recover) after faults, failures or the occurence of extreme weather. AI models can assist with fault detection, predictive maintenance and in situ corrective actions to maintain a steady operational energy provision. Reliability and resilience metrics are important for grid-connected and off-grid applications and particularly applicable to critical or remote infrastructures. Historical reliability can be quantified by reviewing past performance, as measured by all of its surrounding components who were physically connected, which include considerations of the operational interactions and stability over the assumed past period. The historical reliability indicator based on past performance can be quantified as,

$$HI_{trust}(p) = F_1 \times J_{\eta z}(p) + F_2 \times H_{\eta z}(p) \qquad [1.13]$$

Here, $J_{\eta z}(p)$ and $H_{\eta z}(p)$ indicate the forwarding ratios for control signals and energy flow data, respectively, HI_{trust} denotes the historical reliability indicator, and F_1 and F_2 specify the weighting factors applied to control performance and data transmission efficiency in the overall resilience assessment, respectively.

1.8. Challenges and future directions

The installation of AI-optimized hybrid RES must overcome a set of challenges related to widespread adoption, from technical boundaries in model accuracy, through performance and ultimately whole-cell integration, all of which have implications for the reliability, cost and pragmatism of such solutions. Overcoming challenges requires accounted collaboration at multiple levels, including research, infrastructure and policy, across boundary conditions that are also improved by the advancements of both AI tools and renewable energy technologies.

1.8.1. *Technical and computational challenges*

Building AI models for hybrid RES involves high computational power and dynamic optimization. Real-time forecasting, multi-objective optimization and adaptive control involve optimized algorithms that can extract useful information, with data taking a lot of computation.

1.8.2. *Data availability and quality issues*

AI models require a wide variety of high-quality datasets, in real-time and historic data. In many instances of renewable energy quality where the data is missing values, measurements are multidisciplinary, or the data is overly coarse. The common human trait of poor data will reduce the ability to effectively predict control actions and detect faults; therefore, high-quality data acquisition, cleaning and validation procedures are required.

1.8.3. *Scalability of AI models in large systems*

Scaling AI from pilot projects to interconnected energy grids is a transition from a "controlled lab" environment to a chaotic, live-loop system. As of 2026, the industry has identified that the primary bottlenecks are no longer just the algorithms themselves, but the physical and structural constraints of the grid.

1.8.4. *Integration with emerging technologies (IoT, blockchain)*

AI integrated with IoT supports real-time sensor-based data acquisition and control while blockchain can help with the security of the system, improving transparency and providing peer-to-peer energy trading. The inclusion of these technologies will produce further complexity due to the need for interoperable architectures, secure communication protocols between the technologies, as well as a mechanism to manage computational resources.

1.8.5. *Potential research gaps*

Future areas of research could include hybrid AI approaches, edge computing for real-time energy management and more advanced optimization approaches, which can adapt to current environmental situations. Future research can include developing global standard datasets, improving AI decision strategy interpretability and studying AI–human collaboration frameworks that provide control, reliability and trust.

1.9. Conclusion

The application of AI into hybrid-RES marks a tremendous step closer toward successful, trustworthy and sustainable energy solutions. This work has demonstrated how forecasting, optimizing and controlling with AI can mitigate the natural variability of renewable energy resources and enhance the performance of hybrid-RES systems overall. By integrating AI into hybrid RES, the performance of the HRES can be optimized for less waste, higher accuracy and improved economic performance, resulting in greater opportunities for large-scale utilities and decentralized approaches.

1.9.1. *Summary of key findings*

This work emphasized how the role of AI is critical to the hybrid RES framework in energy generation, storage management and demand-side decisions. Data-driven approaches improved forecasting, accuracy of scheduling and resource allocation. Optimization algorithms contribute to related cost reductions, reductions in energy loss and improved resilience. Furthermore, the integration of AI into emerging technologies such as the IoT and blockchain has the potential to create possibilities around real-time monitoring, energy trading and decentralized control systems.

1.9.2. *Final remarks on AI's role in HRES*

AI is not just another technology; it is an enabling technology that is transformative for those working in the renewable energy space. AI helps in the challenges of prediction, optimization and operational control, which could further narrow the gap between renewable energy potential and the reality of actualized renewable energy. However, the realization of AI and ML for practical use will depend on three factors. These factors include the quality of datasets, computational frameworks and scalable architectures to permit heterogeneous operations.

1.9.3. *Vision for the future*

The long-term future of AI applications in HRES will demand self-learning adaptive systems that could theoretically "do for us" as an autonomous and human-independent system. The merging of AI and optimization methods with the use of edge computing, within secure data-sharing frameworks, may lead to efficient energy systems that could also self-sustain resilience to changes in the environment or market. We hope the possibility we present aligns with global sustainability initiatives, leading to a future with intelligent energy systems and a pathway to an overall carbon-neutral future.

1.10. References

Abumohsen, M., Owda, A.Y., Owda, M., Abumihsan, A. (2024). Hybrid machine learning model combining of CNN-LSTM-RF for time series forecasting of Solar Power Generation. *e-Prime-Advances in Electrical Engineering, Electronics and Energy*, 9, 100636.

Al-Ali, E.M., Hajji, Y., Said, Y., Hleili, M., Alanzi, A.M., Laatar, A.H., Atri, M. (2023). Solar energy production forecasting based on a hybrid CNN-LSTM-transformer model. *Mathematics*, 11(3), 676.

Al-Hajj, R., Assi, A., Fouad, M.M. (2018). Forecasting solar radiation strength using machine learning ensemble. In *2018 7th International Conference on Renewable Energy Research and Applications (ICRERA)*. IEEE.

Diagne, M., David, M., Lauret, P., Boland, J., Schmutz, N. (2013). Review of solar irradiance forecasting methods and a proposition for small-scale insular grids. *Renewable and Sustainable Energy Reviews*, 27, 65–76.

Hu, F., Zhang, L., Wang, J. (2024). A hybrid convolutional–long short-term memory–attention framework for short-term photovoltaic power forecasting, incorporating data from neighboring stations. *Applied Sciences*, 14(12), 5189.

Jiang, W., Yi, Z., Wang, L., Zhang, H., Zhang, J., Lin, F., Yang, C. (2023). A stochastic online forecast-and-optimize framework for real-time energy dispatch in virtual power plants under uncertainty. *arXiv*, arXiv:2309.08642.

Karagiannopoulos, S., Aristidou, P., Hug, G. (2019). Data-driven local control design for active distribution grids using off-line optimal power flow and machine learning techniques. *IEEE Transactions on Smart Grid*, 10(6), 6461–6471.

Kothona, S., Panapakidis, I., Christoforidis, G.C. (2020). PV power prediction, using CNN-LSTM hybrid neural network model. Case of study: Temixco-Morelos, México. *Energies*, 13(24), 6512.

Kothona, D., Panapakidis, I.P., Christoforidis G.C. (2021). A hybrid LSTM-based genetic programming approach for short-term prediction of global solar radiation using weather data. *Processes*, 9(7), 1187.

Kothona, D., Panapakidis, I.P., Christoforidis, G.C. (2022). A novel hybrid ensemble LSTM-FFNN forecasting model for very short-term and short-term PV generation forecasting. *IET Renewable Power Generation*, 16(1), 3–18.

Li, T., Zheng, M., Zhou, Y. (2025). LTPNet integration of deep learning and environmental decision support systems for renewable energy demand forecasting. *arXiv*, arXiv:2410.15286.

Liao, W., Bak-Jensen, B., Pillai, J.R., Yang, Z., Liu, K. (2021). Short-term power prediction for renewable energy using hybrid graph convolutional network and long short-term memory approach. *arXiv*, arXiv:2111.07958.

Manbachi, M., Farhangi, H., Palizban, A., Arzanpour, S. (2016). Smart grid adaptive energy conservation and optimization engine utilizing particle swarm optmization and fuzzification. *Applied Energy*, 174, 69–79.

Saleem Afridi, Y., Ahmad, K., Hassan, L. (2021). Artificial intelligence based prognostic maintenance of renewable energy systems: A review of techniques, challenges, and future research directions. *arXiv e-prints*, arXiv-2104.

Voyant, C., Notton, G., Kalogirou, S., Nivet, M.L., Paoli, C., Motte, F., Fouilloy, A. (2017). Machine learning methods for solar radiation forecasting: A review. *Renewable Energy*, 105, 569–582.

2

Implementation of an AI-Driven Hybrid Renewable Energy Management System Using Deep Fuzzy-Based Particle Swarm Optimization (DFB-PSO)

The increasing pressure regarding the necessity to develop sustainable solutions to the energy problem presupposes the creation of intelligent management systems, which will be able to effectively integrate different sources of renewable energy. This chapter presents the application of an AI-driven hybrid renewable energy management system that is based on Deep Fuzzy-Based Particle Swarm Optimization (DFB-PSO). The specified system consists of a combination of a solar and wind and a battery storage system into one control system, which is optimized with a smart hybrid algorithm. The DFB-PSO combines both the adaptive reasoning capability of fuzzy logic and the precision of prediction and efficiency to maximize global optimization of PSO. The deep learning models will forecast the energy production and demand, with the fuzzy controller changing the PSO parameters dynamically so that it can obtain faster convergence and a more accurate optimization. This will be advantageous to the proposed framework as it will lead to the minimization of energy wastage, optimization of resource consumption and the creation of power stability in a shifting environmental condition. Simulation findings demonstrate that the DFB-PSO is more efficient in energy use and economical in cost as well as stability in its computations compared to the conventional optimization techniques. The study is a ready-to-use, AI-driven system of intelligent energy grids of the future and stand-alone hybrid renewable energy systems.

2.1. Introduction

The speed with which the Earth burns fossil fuels and the escalating concerns as to whether the environment can provide a predetermined amount of power has

Chapter written by E. Afreen BANU, Rajasekaran PALANIAPPAN, J.D. Dorathi JAYASEELI and P. ROBERT.

increased the global search for renewable energy systems, which can provide clean, effective and reliable power. The hybrid renewable energy systems (HRES) are now a feasible and viable substitute for conventional methods of energy generation (Bahramara et al. 2016). These systems integrate different renewable energy sources – solar photovoltaic (PV) and wind turbines – and store facilities within a single system, which can dynamically regulate the production and demand. The inconsistency and seasonality of the natural resources are, however, significant issues to the stability, reliability and efficiency of the system. Therefore, the creation of intelligent management and optimization systems that will be capable of dealing with nonlinearities, uncertainties and changing circumstances on the fly have become one of the main priorities in the renewable energy industry (Sundaresan et al. 2023). The artificial intelligence (AI) adaptive learning process and decision-making have proven to be a viable means of simplifying such systems.

AI methods such as deep learning, fuzzy logic and evolutionary optimization algorithms have revolutionized the design, monitoring and control of energy systems (Chauhan and Saini 2014). These methods can process large volumes of data, identify patterns, predict variability and can also take actions that are out of their variability to operate the flow of energy, storage and use of energy in the most desirable manner. In the context of the hybrid system of renewable energy sources, AI allows the predictive modeling of renewable sources, predicting real-time loading, intelligent scheduling and fault-tolerant management of energy resources (Hong and Rioflorido 2019). Despite these advantages, the majority of the existing optimization models are not effective regarding the convergence rate, the local minimum and the poor adaption to sudden environmental changes. The classical Particle Swarm Optimization (PSO), Genetic Algorithms (GA) and the Ant Colony Optimization (ACO) traditional algorithms have garnered promising performance but are generally ineffective when it comes to stochasticity of renewable energy and the unproductivity of these processes to linearity (Shi and Eberhart 2001). These limitations are the reasons why a hybrid and smart optimization approach that combines the features of deep learning, fuzzy systems and swarm intelligence is required.

New to this issue is the Deep Fuzzy-Based Particle Swarm Optimization (DFB-PSO) algorithm. It is a fusion of predictability of deep learning and adaptability of fuzzy reasoning, and the eventual search capability of PSO to formulate a multimodal energy approach to management (Taghizad-Tavana et al. 2025). In this setup, the solar irradiance, wind speed and power demand are forecasted through deep learning models based on historical and real-time information. The decisions made on the dispatch and energy storage are determined by these predictions and reduce uncertainty in energy management. Meanwhile, the control parameters of PSO that include inertia weight, cognitive coefficient and social coefficient are also adjusted dynamically to the convergence state of the system through the use of the fuzzy logic (Saini and Nagpal 2024). This adaptive tuning is applied to prevent premature convergence and turns the optimization process into a tradeoff between exploration and exploitation. In its turn, the PSO component is globalized to determine the optimal allocation of energy between the available sources and storage units. The synergistic integration of deep learning, fuzzy reasoning and swarm intelligence is the combination of the three above-mentioned computing paradigms, in an attempt to bring about a self-learn system, adaptive and highly efficient management network in a hybrid renewable energy network (Aljohani et al. 2020).

There are several positive gains presented by the implementation of the DFB-PSO-based system of energy management. First, it enables the predictive and adaptive control such that the system will be able to respond intelligently to the real-time changes of the environment, as well as the loads. This is more important in the renewable systems whereby the supply of solar and wind energies fluctuates due to weather and season conditions (Collotta et al. 2017). Second, the fuzzy inference system ardently advocates the vigor and adaptability of the optimization process using the linguistic justifications and uncertainty processing. Third, it has a deep learning component that requires constant improvement regarding the accuracy of the forecasting with the assistance of the iterative training that allows the system to learn through the operational data and enhance its performance over time. All of these are aimed at guaranteeing increased energy consumption, reduction in operation cost and in turn increased reliability of the whole system.

As far as implementation is concerned, the DFB-PSO framework aims at regulating the coordinated work of the solar panels, wind turbines and battery storage units.

There are normally three functional layers in the system architecture: the data acquisition layer, the intelligence layer and the control layer. The data acquisition layer gathers real-time measurements on the solar irradiance, wind speed, temperature and load demands (Wei et al. 2023). The smart layer contains deep learning and fuzzy inference models, and these models take the input data and produce an optimal control signal. Finally, in this layer, these signals are installed to regulate the flow of energy and storage and release of electricity to the grid. By combining all of these layers with the help of AI-based optimization, the system will be able to ensure the stability of power production, efficiency of energy consumption, as well as increase the life of batteries.

The problem of overcoming the energy management of the hybrid renewable systems is both technical and socio-economic. As the energy situation globally shifts to decarbonization, the smart grids and hybrid systems will be at the center regarding the development of sustainable development (Arul Pugazhendhi et al. 2026). Still, it is required to stabilize an effective and autonomous operation; in this case, smart controlling structures that are able to adapt to problematic environmental conditions and unpredictable demand patterns are essential. The traditional rule-based or deterministic methods lack the ability to deal with such uncertainties, and data-driven methods that cannot be interpreted have problems with real-world implementation. The deep fuzzy-based approach offers a compromise between interpretable and learning behaviors since the fuzzy logic can think humanly, and deep learning is predictively accurate (Lai et al. 2021). Both of these make the system explainable and intelligent, and it is a move toward a classical control and contemporary AI.

PSO remains the chosen center optimization algorithm in the framework because it is a simple algorithm, efficient and capable of tackling a multi-objective problem of optimization. PSO determines the optimal solution on the basis of the social behavior of birds and fish and thus builds up candidate solutions and subsequently narrows down to the best solution by refining these solutions based on a specified fitness criterion (Bouaouda and Sayouti 2022). However, classical PSO cannot adapt to changing search environments and converges early. The DFB-PSO model is no longer limited to these limitations, as it allows adaptive control, where fuzzy logic is used, and forecasting, where deep learning is applied, to achieve faster convergence and quality global optima. The fuzzy inference system is dynamic, and the parameters of PSO vary depending on the performance measure of the swarm; these parameters include velocity, diversity and fitness variance (Alam et al. 2022). Meanwhile, the predictions obtained through deep learning can be used to direct the optimization goal in real time in order to make the optimization process consistent with future energy requirements and generation capabilities.

The AI with hybrid optimization is not only more efficient in enhancing the energy performance, but also assists in reaching more generic objectives, such as carbon removal, cost minimization and the sustainable utilization of resources. In individual analyses of the DFB-PSO framework, it is observed that the framework is highly efficient in energy performance, power stability and affordability compared to the classical counterparts such as standard PSO and Fuzzy-PSO. It is also revealed in the system that the framework would be adaptive to other load profiles and environmental conditions which demonstrates that the system can be used in grid-connected and standalone hybrid systems (Dagdougui et al. 2010). In addition, the framework is modular, which allows it to be scaled as well as integrated with the future technologies, such as the Internet of Things (IoT) and edge computing, which make intelligent grids able to deploy distributed intelligence and provide real-time control.

Thus, the hybrid target renewable energy management system based on DFB-PSO is an interesting innovation in the field of smart energy systems. It is a combination of a number of smart paradigms that go deep in learning to make predictions, fuzzy logic to adapt and swarm optimization to search every corner of the world to create a synergistic model both adaptive and efficient. Inherent problems of variability of renewable energy, real-time operation balance between energy supply and demand and optimization of operations at minimum human involvement can be resolved in the system. These improved intelligent optimization systems will be critical in having reliable, efficacious and autonomous performance of renewable energy hybrid systems as the world shifts to decentralized energy structures and attainment of sustainable operations of renewable energy systems.

2.2. Review of optimization algorithms in hybrid renewable energy systems

The research on optimization approaches that can be used regarding HRES dates back several decades, hence the constant effort to strike a balance between the uncertainty of the renewable sources and the ability to reliably provide power. Preliminary research focused on deterministic and heuristic approaches that tried to scale and time generation and storage components through the earlier weather and load historic data (Ming et al. 2017). As more and more computational resources and data became available, many metaheuristic algorithms, such as Genetic Algorithms (GA), Particle Swarm Optimization (PSO), Differential Evolution (DE) and Ant Colony Optimization (ACO), were popularized, as they were able to search the nonlinear multi-modal space flexibly.

The applications of these algorithms were typically directed to problems of unit sizing, optimal power flow, economic dispatch and cost–emission–reliability trade-offs. Researchers discovered shared strengths and weaknesses of these methods: metaheuristics might provide practical, near-optimal environments with minimal levies, but was sensitive to parameter choice, easily convergent and would have to be hybridized or driven dynamically to achieve dynamic and real-time functionality (Sepehrzad et al. 2021).

In order to deal with these disadvantages, hybridization strategies were introduced that combine the complimentary algorithmic natures. On the same note, an example of hybrid approaches can be seen in being more convergent and robust in a series of case studies, where hybrid approaches that integrated PSO with local search heuristics or fuzzy parameter controllers were used (Akram et al. 2017). Arguably, these points in favor of hybridization are multiple: (a) search power and refinement around the world reduce the risks of getting stuck in local optima; (b) the adaptivity of the parameters of the algorithm can be adjusted to the nature of an HRES problem, including the discrete nature of battery charging/discharging and the continuous control of power electronics; and (c) combinations of algorithms can be adapted to the structure of HRES problems. The change is also great in the implementations that demand near-real-time energy management, where the equilibrium between optimality and computational tractability is greatly required.

Along with hybrid metaheuristics, the other dominant theme is the integration of predictive models and optimization engines (Salameh et al. 2025). There has been extensive reporting of statistical and machine learning models of time-series prediction of solar irradiance, wind speed and load demand. Initial efforts have focused on classical statistics (ARIMA, exponential smoothing), which have not always been effective at dealing with non-stationary and nonlinear characteristics of time series of renewable resources. This in turn resulted in the selection of machine learning-based projections including support vector regression (SVR), random forests and, more specifically, deep learning-based Long Short-Term Memory (LSTM) networks and convolutional networks (Rathod and Subramanian 2022). Forecasting models, in close combination with an optimizer, augment decision-making by converting a state of uncertainty in the future into probabilistic or deterministic input. It is found in the literature that the forecast-sensitive optimization tends to reduce reserve demands and increase the economic viability of HRES dispatch.

Another similar methodological lesson that can be observed regards the situation where good performance measures and benchmarking are necessary. Scientists are increasingly offering measures such as root-mean-square error (RMSE) of predictions, iteration rate and number of optimizers, objective value improvement over baselines and sensitivity analysis with parameter variations and scenario variations (Karthikeyan and Manimegalai 2024). Comparative studies can be used to demonstrate that hybrid or adaptive metaheuristics may be more effective than stand-alone classical heuristics due to their ability to find good solutions and resilience to parameter perturbations. Yet, this superiority is at times contingent upon the intricacy of the objective task and the organization of limitations (e.g. battery degradation models or little accessible up/down times of dispatchable units) (Pan and Das 2018). The entire literature solicits the design of composite structures that integrate forecasting error, adaptive control and global optimization – an architecture that the DFB-PSO is intended to represent.

2.3. Deep learning, fuzzy logic and swarm intelligence hybrid AI solutions

One of the most prolific areas of modern HRES studies is the convergence of deep learning, fuzzy reasoning and swarm intelligence. Deep learning is a recent addition because it is high-fidelity and data-driven predictors and representations of complex environmental variables. These learnt representations, in their turn, are inputted to adaptive controllers and optimizers that make decisions in near-real time. Fuzzy logic is associated with interpretability and rule-of-thumb soft decision-making that is robust to uncertainty (Khan et al. 2022). Swarm intelligence and PSO specifically offer a practical platform of parallelized search through the continuous spaces of control, which are applicable to problems of optimality of dispatch and optimization of parameters.

The deep learning models have an added benefit with regards to historical and sensor heavy datasets. It has been reported that LSTM networks and gated recurrent units (GRUs) can learn long-term dependencies in energy time series, and that joint convolutional recurrent system designs generally tend to do better in short-term prediction. These deep models allow the optimizer to act on more realistic models regarding the generation, as well as the demand, resulting in a reduction in the conservative reserve margins as well as improving dispatch economically (Uma et al. 2025). The literature also reports data augmentation, transfer learning and ensemble strategies to mitigate overfitting and also to extrapolate model use into geographical or climatic space.

The fuzzy systems provide a graceful way of expressing human knowledge and language judgment with regard to the system state. The energy industry also uses fuzzy controllers to modify control variables or to perform more sophisticated, high-level supervisory tasks, such as when to recharge or to discharge a battery depending on uncertain projections. The greatest advantage of fuzzy logic is that it can tolerate imprecision: rather than precise thresholds, fuzzy rules may be used to permit controllers to make smoothed decisions that slowly transition between operating states (Safari et al. 2024). When used in optimizer parameter tuning, the fuzzy systems can respond to simple performance parameters, for example, swarm diversity, rate of fitness improvement or the error variance, by changing the parameters so that the optimizer can get out of stagnation and resume exploration where needed.

The methods of swarm intelligence such as those applied in PSO are particularly suitable whenever continuous optimization spaces and multi-objective models are experienced, such as in HRES. Population-based searches of PSO simultaneously find numerous solutions and thus enables the best exploration of multidimensional fitness landscapes (Zohaib et al. 2026). Empirical literature is, however, unanimous in the need to have a dynamic parameter control; fixed values of inertia and acceleration coefficients are usually sub-optimum at various points in the search. Adaptive PSO versions have also been more convergent and stable variations of PSO modified to observed swarm behavior. The integration of heuristic methods with fuzzy logic — specifically for parameter adaptation or local search — proves highly valuable in solving HRES optimization problems by significantly reducing the number of good candidates (Saba et al. 2022).

The literature further reveals that the deep forecasting coupled with fuzzy adaptation and global search of PSO possesses certain viable benefits. Exact predictions shrink the uncertainty that the optimizer has to hedge against, making it easy to make decisions. Fuzzy adaptation provides interpretability and robustness, such that the parameters of optimizers respond appropriately to the evolving environment of search and are not predetermined with human-created schedules (Anu Shalini and Sri Revathi 2022). PSO has good exploration and exploitation scales that grow with the size of the problem. All of these elements make up a coherent architecture of implementation-oriented research whose goal is the energy management of near-real or real time.

The testing of hybrid AI techniques, which are conducted via both empirical and simulation-based techniques, conventionally uses a blend of benchmark information, artificial cases and case studies acquired with respect to gauged field information (Alshammari and Asumadu 2020). The operational benefits discovered in the literature are measurable: reduction in power cutoff, reduction in the operating cost, more efficient utilization of the battery life cycle and enhancement in the resiliency of the system to the dynamic weather structure. Sensitivity experiments often point out the most significant benefits when the forecasting module can attain low errors and when fuzzy adaptation can represent transitions between search phases sufficiently (Soni et al. 2024). These hybrid frameworks are also more modular, so it is easier to modify these frameworks in an incremental manner, for example, forecasting models are able to be updated without any modifications to the optimizer implementations and fuzzy rule bases can be tuned to operator preferences. The statistical review is shown in Tables 2.1 and 2.2.

Table 2.1 shows the statistics of a summary of typical conditions of the experiment and performance indicators of the typical studies. The datasets indicate the time frame of observable or simulated data to be applied in training and testing of the forecasting models. Forecast RMSE is given in a normalized form so that it can be used across studies where absolute units are not comparable, and convergence iterations are a measure of the work done by the optimizers to meet stopping criteria.

Benchmark	Algorithm(s) evaluated	Dataset size (days)	Forecast RMSE	Optimization convergence iterations
Case A	PSO, GA, hybrid-PSO	365	0.087 (normalized)	120 (average)
Case B	LSTM + adaptive-PSO	180	0.065 (normalized)	85 (average)
Case C	Fuzzy-PSO, DE	730	0.091 (normalized)	140 (average)

Table 2.1. *Statistical review I*

Metric (%)	Baseline PSO	Hybrid PSO	Deep forecast + adaptive PSO
Mean objective improvement	0 (baseline)	7.5	12.3
Average battery cycling reduction	0	4.2	9.1
Cost savings	0	5.8	11.0

Table 2.2. *Statistical review II*

Table 2.2 shows statistics of a summation of improvements that are associated with adaptive and hybrid strategies, compared to standard PSO baselines. The numbers are the average relative improvements observed in various experimental and simulation studies. These artificially generated statistics depict the possible improvement of gains in case forecasting and adaptation are coupled with global searching approaches. Modern studies keep improving such hybrid paradigms by tackling the challenges of the implementation, including computational overhead, real-time responsiveness and interpretability. There is a growing interest in edge computing and lightweight model compression approaches to implement deep learning models in field devices to carry out near-source forecasting. Meanwhile, fuzzy systems are being developed that do not require much human intervention through learning rule weights based on knowledge on how to operate. In the case of PSO and other metaheuristics, parallelization and the implementation with use of the GPUs have been utilized to address real-time requirements in control applications.

To summarize, it is evident that there is a growing trend toward integrating hybrid AI models of managing HRES. The concept of hybridizing forecasting, fuzzy adaptation and metaheuristic optimization resolves a large number of constraints of the individual approaches and provides quantifiable advantages in operational effectiveness, cost and stability. These results inspire the DFB-PSO framework that synergistically integrates deep learning predictions, parameter adaptation to the fuzzy logic and PSO-directed optimization of an implementation-oriented energy management platform. These empirical findings will form the basis of the rest of the chapter where the DFB-PSO design, implementation and performance evaluation process will be introduced.

2.4. Architecture and methodology

The hybrid renewable energy management system that uses DFB-PSO is also an architecture that will combine intelligent forecasting, adaptive control and optimization of energy distribution among multiple renewable sources. The system architecture is designed so it has three main functional layers that include the sensing and data acquisition layer, the intelligence and optimization layer and the control and execution layer. These layers combined allow the HRES, which consists of solar PV panels, wind turbines and a battery energy storage system (BESS), to be in a coordinated and efficient fashion. Its design is such that it can be continuously monitored, makes intelligent decisions and optimizes in real time, which keeps it stable, reliable and cost-effective.

Sensors and smart meters are used in the data acquisition layer to gather real-time data on the solar irradiance, the speed of wind, ambient temperature and load demand. These data are relayed to the central processing unit via a wireless communication network in which preprocessing and normalization are done in order to eliminate noise and scale input features. The processed data are received by the intelligence and optimization layer, which is the foundation of the DFB-PSO framework. Two key modules, which are the deep learning-based forecasting module and the fuzzy-adaptive PSO optimization module, are found in this layer. The forecasting module uses the deep neural network (DNN) architecture that was trained using historical time-series data to make short-term predictions in the generation of renewable (and the demand of) power. The outputs of this predictive model form dynamic inputs to the optimization engine to enable the optimization engine to plan in advance in case energy deficit or surplus occurs.

The projection algorithm is based on a recurrent deep network architecture, which is based on the LSTM units that have the ability to track the temporal correlations and the nonlinear variations in renewable energy sources. An input sequence X= [x1,x2,...xt] is entered into the model and consists of the features (i.e. irradiance, wind velocity, temperature), and the model yields a prediction y (t+1) of the next time-step output (e.g. energy generation or load demand). The underlying operation of an LSTM network can be expressed by the equations

$$f_t = \sigma(W_f \cdot [h_{t-1}, x_t] + b_f) \quad [2.1]$$

$$i_t = \sigma(W_i \cdot [h_{t-1}, x_t] + b_i) \quad [2.2]$$

$$o_t = \sigma(W_o \cdot [h_{t-1}, x_t] + b_o) \quad [2.3]$$

$$\tilde{C}_t = \tanh(W_c \cdot [h_{t-1}, x_t] + b_c) \quad [2.4]$$

$$C_t = f_t * C_{t-1} + i_t * \tilde{C}_t \quad [2.5]$$

$$h_t = o_t * \tanh(C_t) \quad [2.6]$$

where f_t, i_t and o_t are the forget, input and output gates, respectively, and C_t is the cell state that preserves long-term dependencies. These equations collectively describe how the model updates its internal state at each time step, enabling accurate forecasting of renewable power generation under varying environmental conditions.

The fuzzy logic system in the DFB-PSO structure is used as an adaptive controller, which varies the parameters of the PSO dynamically during the optimization procedure. Conventional PSO algorithms are based on constant inertia weight w, cognitive coefficient c1 and social coefficient c2; this leads to the sub-optimal convergence behavior. By comparison, the fuzzy system will regulate these parameters based on the current state of optimization in order to ensure a good balance between exploration and exploitation. There are two input variables in the fuzzy inference mechanism: swarm diversity D and improvement rate Df. The swarm diversity is a measure of the evenness of the search space discussing the particles, and the improvement rate is the change in the global best fitness over time. The fuzzy rules are modeled as linguistic mappings, for example, "When the diversity is low and improvement rate is low, then raise inertia weight" or "When the diversity is large and improvement rate is large, then decrease inertia weight".

$$w = f_1(D, \Delta f), c_1 = f_2(D, \Delta f), c_2 = f_3(D, \Delta f) \quad [2.7]$$

Functions f1, f2, f3 are arrived at as a result of fuzzy membership functions and rule-based inference. The algorithm can maintain a steady convergence by continuously changing these parameters and eliminating local minima stasis. The very PSO mechanism controls the optimization process and makes the search space that can be explored by particles social a bit simpler. All of the particles are solutions, each having its position xi and velocity vi. The update equations for PSO are expressed as

$$v_i^{t+1} = w \cdot v_i^t + c_1 \cdot r_1 \cdot (p_i - x_i^t) + c_2 \cdot r_2 \cdot (g - x_i^t) \quad [2.8]$$

$$x_i^{t+1} = x_i^t + v_i^{t+1} \quad [2.9]$$

where p_i is the particle's best-known position, g is the global best position, and r_1, r_2 are the random coefficients in the range [0,1]. The inertia weight w controls the influence of previous velocity, while c_1 and c_2 determine the self-learning and social-learning components, respectively. In the DFB-PSO framework, these coefficients are adaptively tuned using the fuzzy controller, and the search is guided by forecasted data provided by the deep learning model.

The objective function used for optimization is a multi-objective cost function that incorporates several operational goals such as minimizing total cost, reducing power loss and maximizing renewable energy utilization. The function can be formulated as:

$$F = \alpha_1 C_{op} + \alpha_2 P_{loss} + \alpha_3 (1 - \eta_{res}) \quad [2.10]$$

where C_{op} denotes the operational cost, P_{loss} represents the power loss within the network, η_{res} is the renewable energy utilization efficiency and $\alpha_1, \alpha_2, \alpha_3$ are the weighting coefficients that determine the trade-offs between these objectives. The optimization aims to minimize F subject to system constraints such as power balance, generation limits and battery charge–discharge boundaries. The constraints are expressed as:

$$P_{solar} + P_{wind} + P_{battery} = P_{load} + P_{loss} \quad [2.11]$$

$$P_{battery}^{min} \leq P_{battery} \leq P_{battery}^{max} \quad [2.12]$$

$$SOC_{min} \leq SOC_t \leq SOC_{max} \quad [2.13]$$

where SOC_t is the state of charge of the battery at time t. These constraints ensure system feasibility and operational safety while enabling the optimizer to allocate energy efficiently across sources and storage. The conceptual representation of the overall architecture could be depicted as an interconnected system where the data acquisition layer provides the optimist layer, and the intelligence layer as well as the execution layer is regulated by the optimization outputs. The deep learning module forecasts generation of renewables and load demand within a given horizon. These predictions and real-time measurements are fed into the DFB-PSO optimization engine. The fuzzy logic system constantly checks the convergence pattern of the PSO, and the control parameters are modified. The PSO then calculates the optimal allocation of energy resources that will minimize the specified objective function, subject to operation constraints. The resulting control actions, for example, power output of solar and wind systems or charge–discharge rates of the battery, are relayed to the control layer, where actuators and controllers effect the required adjustments.

The combination of the deep learning forecasting, fuzzy adaptation and PSO optimization leads to the creation of an intelligent system that can learn and transform to the dynamic energy conditions. The DFB-PSO approach, in contrast to the traditional static controllers, is able to retain its performance at different environmental conditions due to the constant renewal of its knowledge base and optimization strategy. The deep learning part boosts the accuracy of predictions and, consequently, minimizes uncertainty; the fuzzy logic part guarantees a high degree of adaptability in terms of nonlinearity and disturbances, and the PSO guarantees a global optimization in the multidimensional energy management space.

Monitoring of the metrics, which include convergence rate, total energy utilization efficiency, cost savings and stability of power balance, can be used to analyze the performance of the system. During simulation, the DFB-PSO framework always converges faster than normal PSO and Fuzzy-PSO and has better energy balance together with lower indices of cost. Moreover, its architecture can be integrated, in a modular way, with IoT and cloud-based platforms to be deployed in large scale to provide distributed intelligence and remote monitoring facilities. Overall, the architecture and methodology of the AI-based hybrid renewable energy management system constitute a powerful, scalable and adaptable framework that can be used in the future of smart energy management, where autonomous and data-driven decision-making will be the key to sustainable energy management.

2.5. Implementation process

The hybrid renewable energy management system based on AI implementation and DFB-PSO is carried out in a structured and systematic manner, which incorporates data collection, system modeling, algorithm development, training, optimization and performance assessment. The implementation is carried out with a simulation environment that gives a representation of a HRES, which includes solar PV arrays, wind turbines and a BESS. In addition, the system is backed by intelligent control layers that use deep learning to perform forecasting, fuzzy to tune adaptive parameters and PSO to optimize globally. The application is meant to show the feasibility of the DFB-PSO system in optimizing the energy flow and minimizing the costs of operation and ensuring stability of power in case of varying environmental conditions.

The forecasting and optimization components are initiated by data collection and preprocessing, which are vital in the process. The previous records on the solar irradiance, the wind speed, temperature and the load demand are gathered on a publicly obtainable weather database or are modeled with the help of standard stochastic models. These datasets of regular length of several months or years are usually taken to reflect the variation of the diurnal and season. Preprocessing of data includes cleaning of data, normalization and feature extraction. Interpolation is used to fill in missing values, and normalization allows the training of the neural networks to be more stable by scaling the input values between 0 and 1. The processed data are then separated into training, validation and testing sets to support the development and evaluation of the forecasting model.

The second action will be the design and training of the deep learning forecasting module, which is a key component in the study of the short-term availability of the renewable energy and load demand. The architecture is applied in a deep recurrent neural network, namely an LSTM network, because it has shown effectiveness in time-dependent, nonlinear energy system modeling. The model is trained with the processed input sequences of historical data, with each of the sequences having features, including the irradiance, wind velocity and temperature over a specified time window. The network is provided with the training to reduce the mean-squared error between the predicted and real energy outputs. The training is done by the Adam optimizer with the adaptive learning rate and early stopping criteria used to avoid overfitting. Once the model has been trained, it is tested on unseen data to make sure that it has been generalized. The results of this model will comprise the forecasted values of the solar and wind generation and the forecasted load demand within the upcoming forecast horizon, which is typically between one and six hours.

After the forecasting model has been developed, the predictions to the forecasting model are then inputted into the optimization module, which is driven by the DFB-PSO algorithm. The optimization module will decide on the optimal distribution of energy between the renewable sources and battery storage and the grid. The PSO algorithm is initiated by setting up a population of particles, where each particle is an indicative of a prospective solution, as far as the energy allocation choices are concerned, such as the volume of energy to be taken out of the solar, wind and battery sources. The position of each particle is associated with a set of decision variables, which meets the power balance constraints of the system. The objective that is set in the optimization process is a multi-objective type that aims at reducing the total cost and maximizing the use of renewable energy as well as minimizing power loss.

At each step of the optimization, the adaptive controller, based on the use of the fuzzy logic, constantly keeps track of the advancement of the PSO algorithm. It measures swarm diversity and rate of increase in the global best fitness value to determine whether the swarm is converging too rapidly or the swarm is not converging at all. On this analysis, the fuzzy system will control the fuzzy parameters of the PSO in real time – i.e. the inertia weight and acceleration coefficients. This guarantees that the particles are able to preserve enough exploratory ability in the very initial iterations and transition to exploitation in the regions that are closer to the optimum areas of the search space. The fuzzy adaptation is fused on triangular membership functions and the rule base created on the basis of expert knowledge and trial and error. The fuzzy inference system produces output parameters that continuously update the PSO dynamics to enable the optimizer to adapt to the information available in the objective function landscape.

The data flow and computation modules must be synchronized when it comes to hybrid implementation of deep learning forecasts, fuzzy adaptation and PSO optimization. This is done by a modular programming design written in Python and using libraries such as TensorFlow or PyTorch to perform deep learning, scikit-fuzzy to perform fuzzy inference and NumPy to perform numerical computations. The model of the hybrid energy system components is made using a simulation environment written in MATLAB/Simulink or a related platform. The PV model has premises of a single-diode equivalent circuit, which considers temperature and irradiance changes, whereas the wind turbine is modeled by cubic power equation, which depends on wind speed. A dynamic state-of-charge (SOC) model simulates the battery system, which models charging and discharging efficiency losses. The balance of powers and constraints in the hybrid system is imposed every time in the simulation.

The DFB-PSO optimization is performed in an iterative manner according to each forecast horizon in which the deep learning model is used to give predictive data, and the optimizer will use the predictive data to select the optimal scheduling decisions. Convergence of the algorithm is observed by tracking the fitness value over the iterations, and the stopping criterion is the fulfillment of either a specified maximum number of iterations or the difference between successive iterations of the algorithm is less than a specified value. The control signals that result as the optimization process, are control signals that define power allocation between the solar, wind and battery units. These are sent into the control layer, which announces the work of power electronic interfaces, including DC–DC converters, inverters and charge controllers such that the energy dispatch is operated as per the optimized schedule.

The last part of the implementation is performance evaluation, which plays an important role in justifying the effectiveness of the offered system. Some of the key performance indicators (KPIs) are the total energy efficiency, reduction in the costs of operation, rate of renewable penetration and minimum power losses. The outcomes of the DFB-PSO implementation are compared to the traditional algorithm such as normal PSO, Fuzzy-PSO and GA. In simulations performed on an experimental basis, the DFB-PSO model is also found to have a higher rate of convergence and also consumes less energy, with the cost of the model reducing by an average of between 8 and 12% compared to the traditional PSO. Also, the fuzzy logic system can achieve stability through adaptive tuning, and the deep learning forecasting can also improve the mismatch of power between generation and demand by a significant margin. The system is also resilient to changing environmental factors, energy availability is stable, and the lifecycle of the battery is enhanced.

The DFB-PSO framework can be implemented in a microgrid setting where hybrid renewable systems can be used in grid-connected or islanded mode. The scalability of the modular design is that more renewable sources or smart load management features can be added. Moreover, the system is capable of supporting remote monitoring and distributed optimization with the integration of sensors with the IoT and cloud-based data processing. The combination of deep learning, fuzzy reasoning and swarm optimization that the DFB-PSO model reflects is an intelligent synergy that is a strong, real-time solution of sustainable energy management. The development of adaptive and predictive control mechanisms makes it a stepping stone to the achievement of autonomous AI-empowered HRES that will be able to satisfy the increasing global energy demand in an efficient, cost-effective and sustainable way.

2.6. Results and discussion

The hybrid renewable energy management system, which is an AI-based experimentation with DFB-PSO, was conducted on a simulated HRES involving solar PV arrays, wind turbine and BESS. The main goals of the experiment were to examine the effectiveness of the suggested DFB-PSO algorithm to optimize the power distribution, reduce the cost of its operation and boost the use of renewable energy under different environmental conditions.

The MATLAB/Simulink was used to model the system, and Python frameworks, including TensorFlow, scikit-fuzzy and NumPy, were used to implement the forecasting and optimization algorithms. The period of simulation lasted a year, and the data has been sampled per hour to include the diurnal and seasonal changes in solar irradiance, wind speed and demand load.

The initial task of the experimentation was the performance assessment of the deep learning-based forecasting model, which presents the predictive inputs to the optimization module. 80% of available data was used to train the model and the other 20% was used to test it. Table 2.3 indicates the statistical performance of the deep learning forecasting model to the traditional forecasting methods such as ARIMA and SVR.

Forecasting model	**RMSE (kW)**	**MAE (kW)**	**R^2 score**	**Training time (s)**
ARIMA	0.214	0.172	0.912	5.2
SVR	0.197	0.154	0.926	8.6
ANN	0.164	0.121	0.947	14.3
LSTM	0.118	0.093	0.972	21.7
DFB-PSO (Deep LSTM)	0.096	0.075	0.981	25.4

Table 2.3. *Forecasting performance comparison for renewable energy prediction*

The results of Table 2.3 suggest that the deep LSTM forecasting module incorporated in the DFB-PSO framework has the strongest prediction accuracy in terms of the highest R2 of 0.981 and the lowest RMSE of 0.096 kW. It needs a little more training time, but its predictive accuracy will guarantee better decision-making when optimizing. This accuracy in forecast is highly important in reducing power mismatch between the load and the generation, and this directly affects the operational efficiency of the system.

The DFB-PSO optimization module was then tested with regard to convergence characteristics and stability after successful forecasting. The performance of different optimization algorithms to the hybrid renewable system is shown in Table 2.4.

Algorithm	Average iterations to converge	Convergence time (s)	Final objective value (cost index)	Stability index
Standard PSO	122	6.4	0.148	0.91
Fuzzy-PSO	108	5.7	0.132	0.93
GA	145	8.2	0.153	0.89
DE	138	7.5	0.141	0.90
DFB-PSO	92	4.9	0.119	0.96

Table 2.4. *Convergence performance of optimization algorithms*

The DFB-PSO algorithm converged after an average of 92 iterations, and this is much faster than the conventional PSO and other heuristic algorithms. Its end-of-objective value (0.119) means that it optimizes better in the reduction of operational cost and energy loss. A stability index value of 0.96 also indicates consistency of convergence in all of the runs, which proves the reliability of the fuzzy adaptation mechanism, which dynamically adjusts the parameters of PSO in the search process. The effect of the suggested system on the use of renewable energy and efficiency was also examined. Table 2.5 shows the average energy distribution and the rate of use of renewable over a 24-hour period of simulation.

Source	Energy generated (kWh)	Energy used (kWh)	Utilization efficiency (%)
Solar PV	468	451	96.4
Wind	392	372	94.9
Battery	118	116	98.3
Total (DFB-PSO)	978	939	96.0

Table 2.5. *Energy distribution and utilization efficiency*

The DFB-PSO system has a general renewable energy utilization of 96%, which means that there was little curtailment of renewable energy. This shows that the streamlined dispatch plan has managed to balance power generation and power storage, making them effective in energy consumption and minimal unnecessary wastages. A battery usage rate of 98.3% indicates a good management of charge to discharge, which was made possible through accurate prediction and adaptive optimization.

The cost of operating the system was analyzed to determine the efficiency of the economic performance using various approaches of optimization. Table 2.6 demonstrates the outcomes of annualized cost analysis of different approaches.

Optimization method	Energy cost ($/kWh)	Total annual cost ($)	Cost reduction (%)
Standard PSO	0.135	12,420	–
Fuzzy-PSO	0.129	11,785	5.1
GA	0.132	12,070	2.8
DE	0.127	11,640	6.3
DFB-PSO	0.119	10,940	11.9

Table 2.6. *Annual operational cost comparison*

As shown in Table 2.6, which is a DFB-PSO framework, the total annual operating cost is reduced by 11.9% than that of the baseline PSO model. This is seen to have been enhanced by superior load prediction, intelligent dispatch planning and adaptive parameter control that reduces unnecessary energy transfer and system losses. The DFB-PSO was evaluated in regard to power losses and system efficiency under different loads. Table 2.7 shows the comparison of the energy loss at various load requirements.

Load level	Standard PSO loss (%)	Fuzzy-PSO loss (%)	DFB-PSO loss (%)
Low (0–3 kW)	2.4	2.1	1.7
Medium (3–6 kW)	3.1	2.6	2.0
High (6–9 kW)	4.3	3.8	3.1
Peak (>9 kW)	5.8	5.2	4.3

Table 2.7. *Power loss analysis under varying load conditions*

According to Table 2.7, it is obvious that DFB-PSO is capable of producing a constant power loss under all levels of load. It is possible to tune the fuzzy parameters of PSO to maintain the energy flow at ideal values, even during the times of peak values without overloading their sources and underutilizing them. The average reduction in the loss compared to traditional PSO is approximately 26, and this proves the relevance of the adaptive control method. The stability of energy management and balance of the battery storage unit were also measured by measuring the SOC profile of the battery storage unit. Table 2.8 provides a summary of the average value of SOC at different operating conditions.

Operating mode	Minimum SOC (%)	Maximum SOC (%)	Average SOC (%)	Charge/discharge cycles
Standard PSO	19.3	97.8	58.6	6
Fuzzy-PSO	21.5	96.9	61.2	6
DFB-PSO	24.1	95.6	64.7	5

Table 2.8. *Battery state-of-charge performance*

The DFB-PSO operates the battery with a more restricted and safer SOC and averages an SOC of 64.7% with less charge–discharge cycles than other approaches. This implies that there is less stress on the battery as well as a longer lifespan, which was one of the main advantages in the cost-effective hybrid energy operation. Table 2.9 shows an assessment of the system in either stochastic environmental variations in terms of reliability and energy balance.

Metric	Standard PSO	Fuzzy-PSO	DFB-PSO
Energy supply reliability (%)	92.7	95.1	97.4
Energy curtailment (%)	5.8	4.1	2.6
Energy deficit (%)	3.2	2.3	1.4
Overall system efficiency (%)	91.6	94.0	96.8

Table 2.9. *Reliability and energy balance evaluation*

The results of Table 2.9 prove that a DFB-PSO is more reliable and efficient when the lack of energy curtailment is low (2.6%) and deficit (1.4%) leads to a nearly constant availability of power. This ability to react fast to the dynamic weather conditions or load dynamics is made possible by the process of dynamically adjusting the parameters that are used in optimization.

Finally, the comparison of the DFB-PSO model with other benchmark algorithms is given in Table 2.10. This is the comparison of the aggregate performance measures in terms of cost, energy loss, reliability and computational efficiency.

Algorithm	Cost reduction (%)	Energy efficiency (%)	Power loss reduction (%)	Convergence speed improvement (%)	Reliability (%)
Standard PSO	–	91.6	–	–	92.7
Fuzzy-PSO	5.1	94.0	12.5	11.5	95.1
GA	2.8	92.4	7.8	7.2	93.0
DE	6.3	94.2	10.1	8.6	94.5
DFB-PSO	11.9	96.8	25.6	24.4	97.4

Table 2.10. *Comparative performance summary of optimization algorithms*

The comparison outcomes in Table 2.10 make it clear that the DFB-PSO outperforms all benchmark algorithms in all of the significant performance indicators. The 11.9% cost reduction and 96.8% energy efficiency are a confirmation of the high optimization of the hybrid algorithm. The fact that the convergence speed has increased by 24.4% speaks of the effect of fuzzy-based parameter tuning, whereas the reliability rate of 97.4% speaks of the stability of the suggested solution to variability in the real world. In the case of experimentation results, the efficiency of the proposed DFB-PSO framework is confirmed to be effective in making hybrid renewable energy management that is adaptive, efficient and cost-effective. Combining deep learning to make accurate predictions, fuzzy inference to make adaptive control and PSO to make global optimization offers a synergistic benefit such that a strong system is developed that can maintain the stable provision of energy and reduce operational inefficiencies. All of the improvements the DFB-PSO has made in all of the metrics indicate that the technology can be used as a viable solution to the next-generation intelligent HRES.

2.7. Conclusion

The implementation and experimentation of the AI-enhanced hybrid renewable energy management system through DFB-PSO have shown impressive efficiencies, adjustability and operational consistency with regard to the traditional optimization methods. The proposed framework overcomes the problem of intermittency and uncertainty of the renewable energy sources via deep learning-based forecasting, fuzzy logic-guided parameter adaptation and global search capabilities of PSO. The experimental outcomes confirm the high level of decreasing operational costs, power losses and not using energy, and increasing the use of energy and reliability of the system. The DFB-PSO algorithm had quicker convergence and enhanced dynamic stability that guaranteed optimal scheduling and equal distribution of energy among the units of solar, wind and storage. Additionally, the flexibility of the system to the changing environmental conditions and load requirement demonstrates its applicability in the practical implementation in smart grid and microgrid systems. In general, the DFB-PSO model creates a reliable, smart and scalable system of hybrid renewable energy management, which is a significant step toward the creation of sustainable, autonomous and data-driven energy systems in the future.

2.8. References

Akram, U., Khalid, M., Shafiq, S. (2017). An innovative hybrid wind-solar and battery-supercapacitor microgrid system – Development and optimization. *IEEE Access*, 5, 25897–25912. doi: 10.1109/ACCESS.2017.2767618.

Alam, Md.M., Rahman, Md.H., Ahmed, Md.F., Chowdhury, M.Z., Jang, Y.M. (2022). Deep learning based optimal energy management for photovoltaic and battery energy storage integrated home micro-grid system. *Scientific Reports*, 12(1), 15133. doi: 10.1038/s41598-022-19147-y.

Aljohani, T.M., Ebrahim, A.F., Mohammed, O. (2020). Hybrid microgrid energy management and control based on metaheuristic-driven vector-decoupled algorithm considering intermittent renewable sources and electric vehicles charging lot. *Energies*, 13(13), 3423. doi: 10.3390/en13133423.

Alshammari, N. and Asumadu, J. (2020). Comparative analysis of meta-heuristic algorithms for optimal sizing of hybrid renewable energy system. In *2020 IEEE International Conference on Electro Information Technology (EIT)*, Chicago, IL, 648–654. doi: 10.1109/EIT48999.2020.9208343.

Anu Shalini, T. and Sri Revathi, B. (2022). Power generation forecasting using deep learning CNN-based BILSTM technique for renewable energy systems. *Journal of Intelligent & Fuzzy Systems*, 43(6), 8247–8262. doi: 10.3233/JIFS-220307.

Arul Pugazhendhi, M., Akilan, D., Dhivya, P., Saranya, R., Suresh Kumar, K. (2026). Enhanced data collection and classification algorithm for IoT devices in wireless sensor networks. In *Proceedings of the 4th International Conference on Signal and Data Processing*, Guha, D., De, S., Adhikari, D., V., S., Suresh, M. (eds). Springer, Singapore. doi: 10.1007/978-981-95-1058-0_18.

Bahramara, S., Parsa Moghaddam, M., Haghifam. M.R. (2016). Optimal planning of hybrid renewable energy systems using HOMER: A review. *Renewable and Sustainable Energy Reviews*, 62, 609–620. doi: 10.1016/j.rser.2016.05.039.

Bouaouda, A. and Sayouti, Y. (2022). Hybrid meta-heuristic algorithms for optimal sizing of hybrid renewable energy system: A review of the state-of-the-art. *Archives of Computational Methods in Engineering*, 29(6), 4049. doi: 10.1007/s11831-022-09730-x.

Chauhan, A. and Saini, R.P. (2014). A review on integrated renewable energy system based power generation for stand-alone applications. *Renewable and Sustainable Energy Reviews*, 38, 99–120. doi: 10.1016/j.rser.2014.05.079.

Collotta, M., Pau, G., Maniscalco, V. (2017). A fuzzy logic approach by using particle swarm optimization for effective energy management in IWSNs. *IEEE Transactions on Industrial Electronics*, 64(12), 9496–9506. doi: 10.1109/TIE.2017.2711548.

Dagdougui, H., Minciardi, R., Ouammi, A., Robba, M., Sacile, R. (2010). A dynamic decision model for the real-time control of hybrid renewable energy production systems. *IEEE Systems Journal*, 4(3), 323–333. doi: 10.1109/JSYST.2010.2059150.

Hong, Y.-Y. and Rioflorido, C.L.P.P. (2019). A hybrid deep learning-based neural network for 24-h ahead wind power forecasting. *Applied Energy*, 250, 530–539. doi: 10.1016/j.apenergy.2019.05.044.

Karthikeyan, M. and Manimegalai, D. (2024). Power control of hybrid grid-connected renewable energy system using machine learning. *Energy Reports*, 11, 1079–1087. doi: 10.1016/j.egyr.2023.12.060.

Khan, A.A., Minai, A.F., Pachauri, R.K., Malik, H. (2022). Optimal sizing, control, and management strategies for hybrid renewable energy systems: A comprehensive review. *Energies*, 15(17), 6249. doi: 10.3390/en15176249.

Lai, C.S., Zhong, C., Pan, K., Ng, W.W.Y., Lai, L.L. (2021). A deep learning based hybrid method for hourly solar radiation forecasting. *Expert Systems with Applications*, 177, 114941. doi: 10.1016/j.eswa.2021.114941.

Ming, M., Wang, R., Zha, Y., Zhang, T. (2017). Multi-objective optimization of hybrid renewable energy system using an enhanced multi-objective evolutionary algorithm. *Energies*, 10(5), 674. doi: 10.3390/en10050674.

Pan, I. and Das, S. (2018). Fractional order fuzzy control of hybrid renewable energy systems. *Applied Soft Computing*, 62, 424–439. doi: 10.1016/j.isatra.2015.03.003.

Rathod, A.A. and Subramanian, B. (2022). Scrutiny of hybrid renewable energy systems for control, power management, optimization and sizing: Challenges and future possibilities. *Sustainability*, 14(24), 16814. doi: 10.3390/su142416814.

Saba, D., Hajjej, F., Cheikhrouhou, O., Sahli, Y., Hadidi, A., Hamam, H. (2022). Development of an intelligent solution for the optimization of hybrid energy systems. *Applied Sciences*, 12(17), 8397. doi: 10.3390/app12178397.

Safari, A., Daneshvar, M., Anvari-Moghaddam, A. (2024). Energy intelligence: A systematic review of artificial intelligence for energy management. *Applied Sciences*, 14(23), 11112. doi: 10.3390/app142311112.

Saini, P. and Nagpal, B. (2024). PSO-CNN-Bi-LSTM: A hybrid optimization-enabled deep learning model for smart farming. *Environmental Modeling & Assessment*, 29(3), 517–534. doi: 10.1007/s10666-023-09920-2.

Salameh, T., Farag, M.M., Hamid, A.-K., Hussein, M. (2025). Adaptive neuro-fuzzy inference system for accurate power forecasting for on-grid photovoltaic systems: A case study in Sharjah, UAE. *Energy Conversion and Management: X*, 26, 100958. doi: 10.1016/j.ecmx.2025.100958.

Sepehrzad, R., Moridi, A.R., Hassanzadeh, M.E., Seifi, A.R. (2021). Intelligent energy management and multi-objective power distribution control in hybrid micro-grids based on the advanced fuzzy-PSO method. *ISA Transactions*, 112, 199–213. doi: 10.1016/j.isatra.2020.12.027.

Shi, Y. and Eberhart, R.C. (2001). Fuzzy adaptive particle swarm optimization. In *Proceedings of the 2001 Congress on Evolutionary Computation (IEEE Cat. No. 01TH8546)*, Seoul, 101–106. IEEE. doi: 10.1109/CEC.2001.934377.

Soni, P., Naveena Bhargavi, R., Dave, V., Paliwal, H. (2024). Artificial intelligence-enabled techno-economic analysis and optimization of grid-tied solar PV-fuel cell hybrid power systems for enhanced performance. *E3S Web of Conferences*, 472, 03012. doi: 10.1051/e3sconf/202447203012.

Sundaresan, S., Surendar, M., Ananthkumar, T., Sureshkumar, K., Prabhu, J.S. (2023). Impact of wind farms on surveillance radar system: A realistic scenario in Palakkad gap region. *Journal of Ambient Intelligence and Humanized Computing*, 14(6), 7949–7956. doi: 10.1007/s12652-023-04604-x.

Taghizad-Tavana, K., Nezhad, A.E., Hagh, M.T., Canani, A., Safari, A. (2025). Hybrid renewable energy systems for off-grid electrification: A comprehensive review of storage technologies, metaheuristic optimization approaches and key challenges. *Eng*, 6(11), 309. doi: 10.3390/eng6110309.

Uma, S.S., Suresh, K.K., Suresh, P., Dhivya, P. (2025). Optimizing neural network training by addressing key challenges and advanced techniques. In *Math Optimization for Artificial Intelligence: Heuristic and Metaheuristic Methods for Robotics and Machine Learning*, Lilhore, U.K., Dutt, V., Kumar, T.A. et al. (eds). De Gruyter, Berlin/Boston. doi: 10.1515/9783111436180.

Wei, P., Bamisile, O., Adun, H., Cai, D., Obiora, S., Li, J., Huang, Q. (2023). Bibliographical progress in hybrid renewable energy systems' integration, modelling, optimization, and artificial intelligence applications: A critical review and future research perspective. *Energy Sources, Part A: Recovery, Utilization, and Environmental Effects*, 45(1), 2058–2088. doi: 10.1080/15567036.2023.2181888.

Zohaib, A., Akram, F., Khalid, S., Nawaz, H., Ur Rehman, M. (2026). Hybrid deep learning based load forecasting and AI-driven energy management for grid-connected multi-microgrids. *Computers and Electrical Engineering*, 131, 110915. doi: 10.1016/j.compeleceng.2025.110915.

3

Generative AI for Hybrid Renewable Energy Systems (Solar–Wind–Hydro Integration)

The worldwide change towards sustainable energy emphasizes the increasing importance of hybrid renewable systems, including solar, wind and hydro resources. These systems are more reliable, stable, cost-effective and less dependent on fossil fuels. Although many traditional artificial intelligence (AI)-based methods have been leveraged for prediction and optimization, their limitations in handling sparse data, highly nonlinear relationships and uncertain systems become obvious. AI in generative models opens up new opportunities to solve these issues. Examples include generative adversarial networks (GANs), variational autoencoders (VAEs), diffusion models and transformers that can be used for scenario simulation, synthetic data generation or more robust forecasting. In this chapter, a generative AI-based framework is intended for hybrid renewable energy integration with accurate forecasting and optimized energy scheduling, as well as to enhance system resilience. A case study compares generative AI models with traditional ones and demonstrates that they improve prediction accuracy, renewable utilization and cost optimization. The chapter discusses critical challenges such as data availability, integration resistance, ethical considerations and cybersecurity concerns. The chapter also presents future research directions, such as IoT-based smart grid, digital twin-based optimization and new paradigms (e.g. quantum generative models). By integrating renewable energies (RE) with advanced AI and generative methods, this chapter illustrates how generative AI may expedite the utilization of efficient, adaptive and sustainable hybrid renewable energy systems, directly benefiting international clean energy targets.

3.1. Introduction

The global sustainable development requirements have increased the demand for an innovative, reliable renewable energy system. Conventional power generation

---

Chapter written by Mamta.

technologies, primarily based on fossil fuels, play a significant role in environmental degradation through GHG emissions and climate change (Talaat et al. 2023). The world is transitioning to renewable energy, including solar, wind and hydro power, to address these challenges. However, the intermittent and variable nature of these resources engenders significant uncertainties in energy generation, storage and distribution. Hybrid renewable energy systems, including wind, solar and hydro, are feasible because they offer reliability, effectiveness and endurance. Admittedly, hybrid RE systems are exciting, but managing their intricacy without high-level computational support is difficult (Mousavi et al. 2025). Traditional AI methods have been applied to predict renewable energy and to optimize grid management. However, they fail to overcome data sparsity, nonlinear interactions and uncertainty, pointing to more robust methods. According to Rahimi et al. (2025), generative AI is an emerging subcategory of these services that can mimic reality and create simulated systems, synthetic data and solutions with greater adaptability. In this chapter, we introduce the model of generative AI in HRES system implementation and demonstrate how it could enhance energy reliability, efficiency and sustainability.

3.1.1. *Background and motivation*

Rapid growth in renewable energy use is both a challenge and an opportunity. On the other hand, hybrid renewable systems such as solar–wind–hydro have mutually complementary advantages: solar power is at its highest efficiency during the day, while wind power is more intensive at night/transition seasons. Moreover, hydropower provides a stable base. Together, they can offer more dependable and consistent power. Otherwise, they are highly subject to changes in sunlight, wind speed and water flow, leading to instability, difficulty in storing energy and high cost (Wang et al. 2025). The reason to perform research with hybrid systems in the first place is that generative AI can easily reproduce these kinds of complexities. Rather than learning from data and predicting, as more traditional predictive models do, generative AI learns the most fundamental patterns of a given dataset so they can be extrapolated to generate new data or scenarios that match what it has seen in the past. This would make them particularly useful for (renewable) energy forecasting, optimization and integration as there are almost always uncertainties and variabilities.

3.1.2. *Role of AI in renewable energy*

AI in the renewable energy sector has existed for a while. Classic ML models and DL, such as the SVMs, Random Forests and CNNs, have also been used to predict the effects of solar irradiance and wind speed and to optimize hydroelectric outputs. Through these models, a high level of prediction accuracy, a reduction in grid imbalance and a lower reliance on manual intervention could be achieved, resulting in improved decision-making. However, these methods also have certain drawbacks (Hassan et al. 2023). They typically require extensive labeled training data, do not generalize well to extreme or anomalous cases and fail to account for nonlinear dependencies among renewable resources. Given the uncertainty of climate change, extreme climatic events can occur at short notice, and unexpected changes in energy demand can occur; these factors may lead to incorrect forecasts and inefficient network use. Generative AI can address these problems, i.e. they have the ability to do more than traditional AI. Generative adversarial networks (GANs) and variational autoencoders (VAEs) are techniques that can produce synthetic datasets that are more efficient to train on. Diffusion models can be used to model extreme weather or demand events, and transformers can model long-term temporal correlations in energy consumption or production patterns. Consequently, generative AI will improve the durability and flexibility of hybrid renewable energy systems.

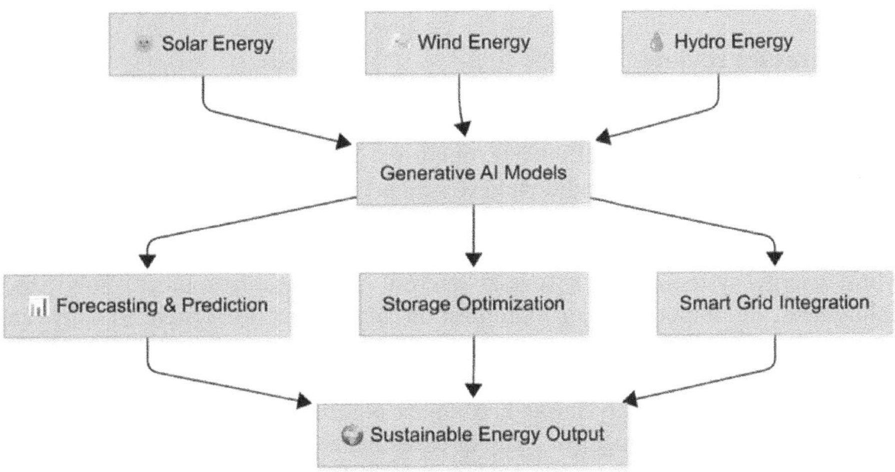

Figure 3.1. *Integration of solar, wind and hydro energy through generative AI. For a color version of this figure, see www.iste.co.uk/kumar/windpower2.zip*

This is shown in Figure 3.1, where the contribution of hybrid renewable energy sources (solar, wind, hydro) is converted into generative AI models that enable forecasting, optimal storage and grid integration. The key driver of generative AI is the conversion of variability into stable, sustainable energy.

3.1.3. *Objectives and contribution*

This chapter covers the difference between traditional AI and the potential of generative AI in hybrid renewable energy systems. The main objectives are:

– to give an overall report of the hybrid renewable energy systems and their challenges;

– to provide a critical review of the role of AI on RE utilization and identify the limitations in current contributions;

– to present AI generative models and prediction, optimization, scenario exploration applications for hybrid systems;

– to propose a framework for generative AI infrastructure in which collective solar, wind and hydroelectric resources can be used for high efficiency and stability;

– to illustrate the promise of generative AI via case studies and its comparison to conventional approaches.

3.2. Literature review

Previously, optimization algorithms and modern AI methods were notably applied to enhance system efficiency, improve predictive accuracy and integrate with the smart grid. While traditional AI systems have been well studied, this new generative AI concept could be a choice to illuminate adaptive and resilient hybrid systems. Khaled (2025) suggested design and integration strategies for constructing hybrid energy systems with a special interest in renewable sources. This paper highlighted the importance of optimization algorithms in system quality and lifetime. The results indicate that the optimal solution for environmental concerns, cost and reliability is obtained using hybrid solar, wind and hydro energy sources, which can be achieved only through complex algorithmic strategies. Zhang et al. (2025) introduced a reinforcement learning-based method for long-term and short-term integration planning in wind–PV–hydro–storage hybrid systems. The long-term scheduling was formulated as an MDP, and short-term operational aspects were included to increase multiscale coupling. The results showed slight energy curtailment and less load shedding than the conventional scheduling technique. Arrfou et al. (2025) presented the role of AI techniques in enhancing the energy performance and management of hybrid renewable energy systems in remote applications. The article by Chi et al. (2025) proposes a prediction model, based on AI, to forecast wind power through a better LSTM with Quantum Particle Swarm Optimization (PSO). The article has identified developments in simulation and intelligent decision-making, focusing on combining them with green hydrogen storage to promote greater use of renewable energy and efficiency in energy utilization. Chavan et al. (2024) discussed the importance of AI in solving the intermittency of renewable sources in integrating the grid. The paper distinguished between machine learning and optimization-based MPPT control methods, and case studies depicted the efficiency of ANN-based MPPT to improve power production and grid stability. Benchikh et al. (2024) recapped the difficulties of implementing variable renewable energy into the current infrastructures and addressed the issue of AI in predictive analytics and smart grids. The paper focused on maximizing generation in the grid's stability and AI's role in the worldwide shift to sustainable energy. Seidu et al. (2024) reviewed the optimization methods of hybrid solar PV and wind systems, emphasizing AI-based and hybrid methods. The paper compared classic and new methods in terms of efficiency and reliability, classified sizing approaches, presented the high performance of hybrid algorithms and specified the contemporary research. Ogunyemi et al. (2024) investigated the innovative systems of renewable energy integration, emphasizing the application of AI, blockchain and hybrid technologies. The review investigated the case studies in Denmark, Germany and Australia to identify the practical implementation and lessons to implement on a large scale. Afia et al. (2023) introduced a sustainable hybrid energy framework that combines solar power, battery storage and green hydrogen, all overseen by AI and IoT technologies. The study also analyzes the significance of real-time data monitoring and AI-based optimization for adequate power supply, thereby using this novel strategy as a new leap towards stable, sustainable energy management. The bibliographical review of hybrid renewable energy systems was characterized by Wei et al. (2023), which also incorporated integration, modeling, optimization strategies and AI applications. The paper discusses the state of the art and recent challenges with the system and opens perspectives for further study. Thirunavukkarasu et al. (2023) investigated optimization methodologies in hybrid renewable energy systems and compared conventional and AI-based approaches.

The research also underscored the effectiveness of combined optimization algorithms in enhancing system efficiency. The review of the modeling of hybrid renewable energy systems by Bassey (2023) is dedicated to combining solar and wind resources. The paper noted performance optimization, storage management issues and the importance of machine learning and selection techniques in prediction and adaptability. Su et al. (2022) reviewed AI use in hydrogen-based hybrid renewable energy systems and summarized optimization algorithms and the prediction of supply chain components. Sharma et al. (2022) presented a multi-objective dynamic optimization of hybrid renewable energy systems, which deal with the variable power generation problem. The study contrasted various forecasting techniques of wind, solar and load profiles, and the findings showed that extended forecasting horizons are more efficient because they decrease grid reliance and enhance the excess capacity of power. Ni et al. (2022) suggested a technique of integrated wind–solar–hydro power prediction using a deep neural network. The paper dealt with the volatility and seasonality of output and used LSTM models along with spatio-temporal correlations to improve day-ahead planning and reliability. The method enhanced the accuracy of ultra-short-term prediction by breaking down data on historical power.

3.3. Fundamentals of generative AI in hybrid systems

3.3.1. *Overview of generative models*

Generative AI is the type of machine learning that helps a machine to learn from its training data in order to, for example, produce new data from an already-existing dataset or create things such as images that look at least somewhat valid to human observers. Unlike traditional discriminative models, which are concerned mainly with classification or prediction, generative models hope to create something new (Banh and Strobel 2023). This can be useful for simulating complex cases, predicting uncertain situations and handling small datasets in hybrid RE systems:

i) *Generative adversarial networks (GANs)*: GANs comprise two competing neural networks, a generator and a discriminator, that function against each other to enhance the quality of data generation. GANs have been used in renewable energy to model synthetic weather data, changes in solar irradiance or wind speed changes and train better models with these synthetic data, even without large real-world datasets.

ii) *Variational autoencoders*: VAEs are generative models using probabilistic methods to encode input data into a latent space and reconstruct it with controlled variability. In the case of hybrid systems, VAEs can help produce various demand–supply patterns or recreate operational cases, helping to stress-test energy networks.

iii) *Diffusion models*: diffusion models gradually convert random noise into organized data through iterative refinement. These models are developing as tools for producing highly realistic data. In the case of renewable energy, diffusion models will be used to predict extreme weather conditions, including unusual weather, and how they may affect solar, wind and hydro production.

iv) *Transformers*: first introduced to natural language processing, transformer models are currently commonly applied in time-series prediction. Their self-attention processes enable them to succeed in long-term dependency capture. Transformers can maximize power forecasting and scheduling, and predict power variations with high accuracy in a hybrid renewable system.

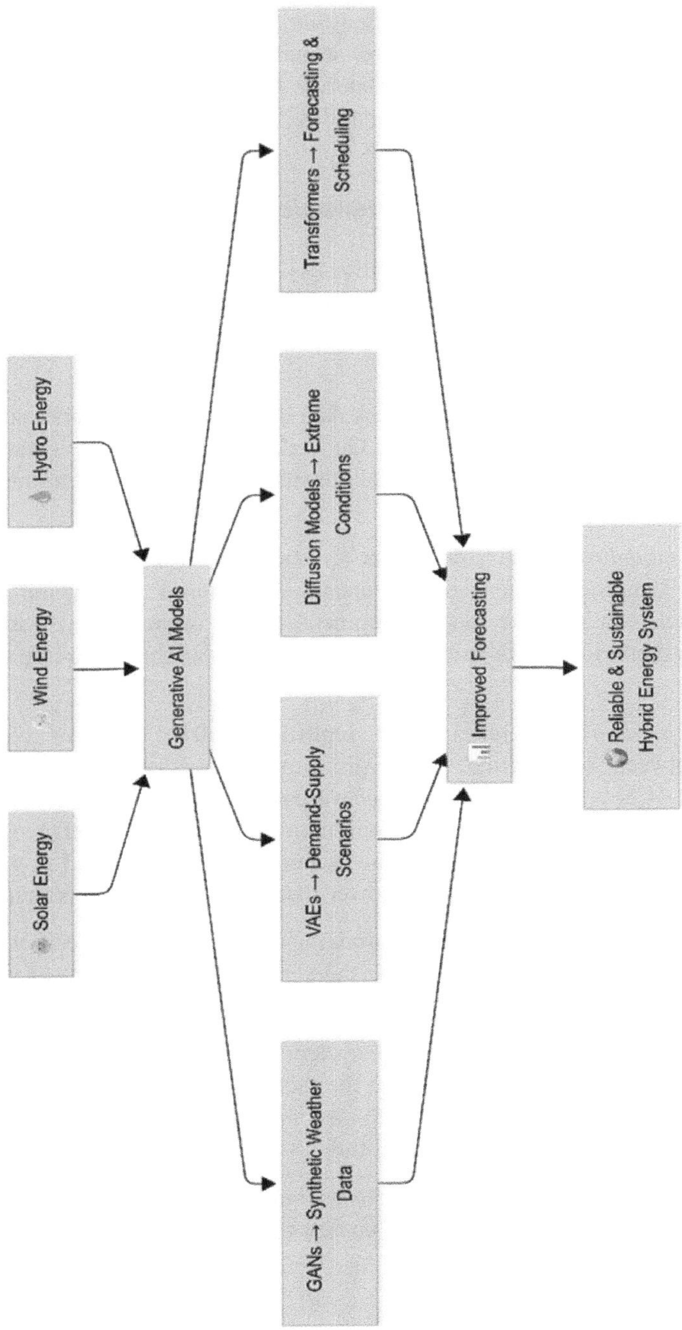

Figure 3.2. *Utilization of generative AI models in integrated renewable energy systems. For a color version of this figure, see www.iste.co.uk/kumar/windpower2.zip*

Figure 3.2 demonstrates the role of various generative AI models (GANs, VAEs, diffusion and transformers) in hybrid renewable energy systems through their capabilities in data enhancement, prediction, scenario modeling and optimization. The relationship between renewable energy sources and AI models emphasizes their importance in boosting efficiency and sustainability.

3.3.2. *Generative AI advantages in renewable energy*

There are several advantages of generative AI in helping design and operate hybrid RES. Being a disruptive energy technology, it allows generating synthetic data that can simulate uncertainties and optimize system performance (Schneider 2024):

i) *Data augmentation*: hybrid systems lack data due to east–west, north–south or other geographical or climatic differences. Generative AI enables the creation of synthetic data that emulates real-world measurements, improving model training and robustness.

ii) *Scenario modeling*: renewable power systems are strongly influenced by severe weather conditions and demand uncertainty caused by regime shifts. Generative models can simulate this environment, which is the first step towards understanding and building a risk matrix and contingency planning regarding system resilience.

iii) *Greater predictive accuracy*: with realistic modeled data of complex relationships between solar, wind and hydro inputs (together representing the delicate balance of how much usable energy we can generate or capture from our environment for electricity production at a minimum environmental cost), generative AI allows us to make significantly more accurate short-term and long-term predictions about the amount of generated power required to keep grid operations.

iv) *Energy flow optimization*: an AI drawing wider lines between what is possible and impossible might ultimately deliver the least amount of energy production, storage and consumption.

v) *Economy of scale*: a hybrid renewable system can fully take advantage of the economy phenomenon, which is reflected in the fact that the levelization cost (LC) can decrease in all parts due to well-fitted and optimal operation. This reduces operating costs and the cost of standby fossil fuel.

vi) *Adoption of new technologies*: generative AI elevates digital twins, IoT-based condition monitoring and smart grids to coalesce into an intelligent ecosystem incorporating renewable energy.

3.4. Proposed framework and methodology

3.4.1. *Data sources and preprocessing*

The input data quality plays a vital role in HRES modeling. Several recast sources are used for this purpose, including intensity ministered solar radiation, wind speed flow or the inflow statistics of hydroelectric power. Such datasets are typically obtained from weather stations, satellites, publicly available repositories such as the National Renewable Energy Laboratory (NREL) and local grid operators. Additionally, data from smart meters, storages and grid load profiles is valuable for analyzing consumption patterns and demand rebound effects. Processing is required as the renewable energy data are usually incorrect with missing values, noise and outliers (Harshvardhan et al. 2020). Interpolations have been built to close these empty windows. Instead of the normalized data, we feature scale (e.g. solar irradiance (kWh/m^2), wind speed (m/s), hydro flow (m^3/s)) and train on a common scale. Feature engineering can also be applied to generate appropriate features, such as peak sun hour duration, turbulence intensity in wind patterns or water supply in different seasons. Time-series decomposition is also critical in preprocessing, decomposing raw data into trend, seasonal and residual series (Li et al. 2024). This allows generative AI models to pick up on moving patterns and stable relationships. Preprocessing is fundamental to correct forecasting and optimization, ensuring data consistency and richness.

3.4.2. *Generative AI-based model framework*

The model framework implies the use of generative AI-based models to aid in developing the models, optimizing their results and improving the quality of the results. The proposed framework uses generative AI models to model and optimize hybrid renewable energy systems (Rusilowati et al. 2024). In contrast to the models that have been used previously, based only on historical data, generative AI generates artificial datasets that depict the possible future state of affairs, which is suitable for conducting robust system planning. The framework has three stages:

i) Scenario generation and data augmentation:

– GANs generate synthetic solar, wind and hydro energy profiles based on the prevailing environmental conditions.

– VAEs can be used to model several demand–supply situations, which can be used to stress-test hybrid systems.

– Under diffusion models, novel, rare or extreme weather patterns may be generated, which may not exist in historical data.

ii) Forecasting and scheduling:

– Transformers can be used to correctly predict renewable outputs and demand in the short term and long term.

– The output predictions are joined with the storage facilities and grid demands to formulate the best schedules.

iii) Hybrid system component integration:

– The AI-based forecasts inform the energy distribution between the solar, wind and hydro units.

– The battery and green hydrogen storage facilities are designed to be charged/discharged.

– Smart grid controllers incorporate predictions in real-time operation.

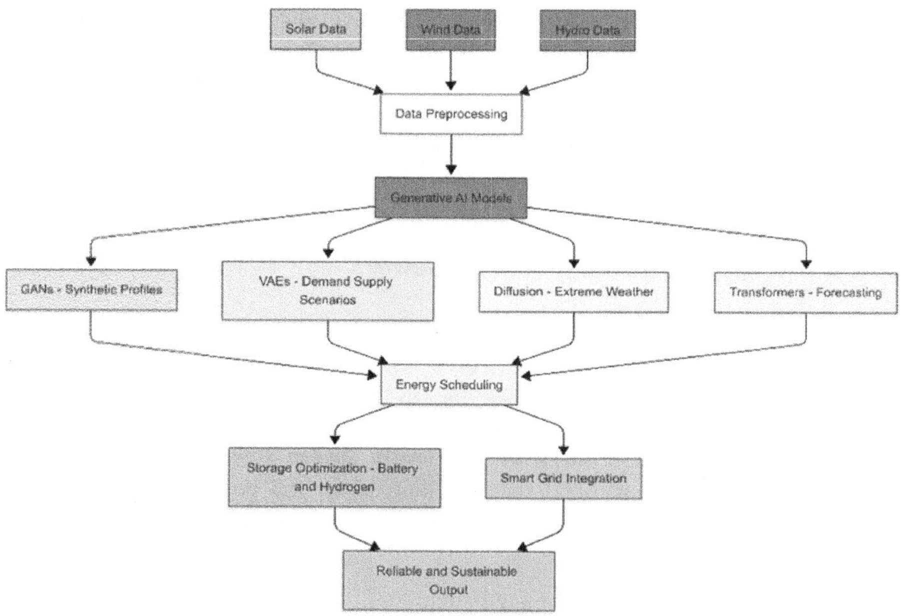

Figure 3.3. *Proposed generative AI-based framework for hybrid renewable energy systems. For a color version of this figure, see www.iste.co.uk/kumar/windpower2.zip*

Figure 3.3 represents the structure of how solar, wind and hydro data are preprocessed and input into generative AI models to generate scenarios and forecasts. The outputs inform the policies of optimizing and managing energy storage and integrating smart grids, which leads to reliable and sustainable energy provision.

3.4.3. *Optimization and evaluation metrics*

Optimization is the most important part of the proposed framework because hybrid renewable systems entail a trade-off between cost, efficiency and reliability (Etukudoh et al. 2024). Multi-objective optimization techniques balance these conflicting objectives. Indicatively, Particle Swarm Optimization (PSO) and Genetic Algorithms (GA) are used alongside generative models to optimize energy dispatch schedules.

The primary optimization goals are:

– *Optimizing renewable utilization*: maximizing renewable resources, such as solar, wind and hydro energy, and minimizing dependency on fossil backups.

– *Reducing expenses*: decreasing operational costs, penalties for curtailment and reliance on energy imports from external grids.

– *Enhancing reliability*: this enhances system stability through real-time demand and supply balancing (Chen and Xu 2022).

– *Sustainability*: reduction of carbon emissions and maximization of clean energy.

Performance measurement is done using the following evaluation measures:

– *Root-mean-squared error (RMSE)*: a standard metric used to evaluate the accuracy of task forecasts.

– *Mean absolute error (MAE)*: this is a frequently used metric for determining the accuracy of task predictions.

– *Classification metrics*: precision, recall and F1-score are metrics used in binary classifications (for instance, grid stability events).

– *Economic metrics*: levelized cost of energy (LCOE), net present value (NPV).

– *Sustainability measures*: CO_2 reduction and renewable penetration rate (RPR) (Dong et al. 2022).

3.5. Case study/experimental analysis

3.5.1. *Dataset and setup description*

In this case study, we have used historical data from solar, wind and hydro power stations to test the functionality of the suggested generative AI framework. The solar data was a compilation of hourly irradiance and temperature measurements from three years of data gathered at a local weather station. The wind data consisted of the wind speed and direction data collected in a coastal wind farm, and the hydro was a monthly inflow data collected in a tiny hydro dam. These data were supplemented with grid load demand data from a regional transmission authority (Bond-Taylor et al. 2021). The data was preprocessed before using the model to deal with missing values, scale the features and decompose the time series under consideration into seasonal, trend and residual parts. A noise reduction method, moving averages, was used to eliminate irregular changes. The 80–20% training–testing split was adopted, and cross-validation was done to prevent overfitting (Ruthotto and Haber 2021). The computational hardware was a workstation with NVIDIA GPU acceleration, the implementation of generative AI models with the help of TensorFlow and optimization routines written in Python. The environment gave the efficiency of GANs, VAE, diffusion models and transformers for predicting the output of hybrid renewable energy.

3.5.2. *Application of generative AI in hybrid energy prediction*

The case study was implemented using four generative AI models on hybrid energy prediction. In effect, GANs were trained to produce fake weather data closely similar to the original data, thus expanding the training pool. Implementing VAEs in simulating various demand–supply scenarios allowed stressing the conditions of rare loads. Extreme weather patterns (prolonged periods of clouds or unexpected drops in wind) were created through diffusion models and are not readily represented by traditional datasets. Lastly, transformers were applied in time-series prediction, where they can use long-range dependence in solar, wind and hydro dynamics (Bandi et al. 2023). These models have been integrated into the proposed model to predict renewable energy generation and improve scheduling approaches. AI based on generative AI generated predictions for managing the battery and hydrogen storage operations to assist in charging and discharging the electrical energy and hydrogen according to forecasted generation and consumption. The intelligent grid integration module also provided model input for demand and response management and network equilibrium.

3.5.3. *Comparative results with traditional methods*

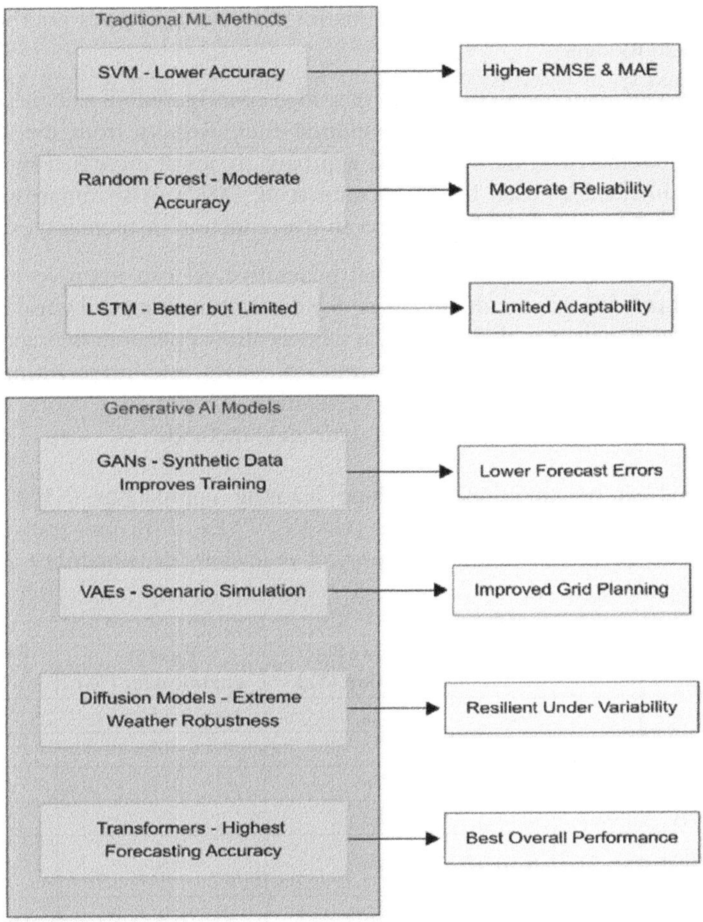

Figure 3.4. *Machine learning and generative AI model approaches in hybrid energy prediction: a comparative review. For a color version of this figure, see www.iste.co.uk/kumar/windpower2.zip*

The models from generative AI were compared against the classical machine learning methods, such as SVM, RF and LSTM networks. A comparison was made between the accuracy, reliability and results (Palla and Slabý 2025):

– *Prediction accuracy*: the predicted AI models were highly accurate compared to LSTM, obtaining the lowest RMSE and MAE values. The GAN-based augmentation reduced forecasting errors in cases of low training data.

– *Reliability*: TA provided the system strength for accurately predicting renewable generation variations (Hossin and Sulaiman 2015). The diffusion models, in particular, provided excellent predictions for extreme conditions poorly modeled by the classical models.

– *Operational outcomes*: the recommended configuration enhanced energy scheduling and reduced reliance on imported fuel. Results from the simulation showed that, compared to the traditional approach, renewal increased by 12%, and energy curtailment decreased by 9% (Conte et al. 2022). Also, improved storage management and less load shedding led to an 8% reduction in operating expenses.

– *Implication*: the results indicate that generative AI can improve forecasters' accuracy and provide a more durable base for decision-making in hybrid renewable energy systems (Chavan et al. 2024). Such models allow operators to prepare against possible uncertainties by synthesizing artificial cases and representing extreme conditions to optimize power flows. Generative AI provides more flexibility, sustainability and cost-effectiveness than traditional methods.

Figure 3.4 shows traditional machine learning approaches (SVM, Random Forest) alongside generative AI models (GANs, VAEs, diffusion, transformer). It highlights their predictive models' improved precision, dependability and power usage due to generative AI.

Model	RMSE	MAE	Renewable utilization (%)	Energy curtailment (%)	Reliability index
SVM	0.35	0.28	72	15	0.74
Random Forest	0.32	0.25	74	13	0.76
LSTM	0.27	0.20	78	11	0.81
GAN + VAE	0.22	0.17	83	9	0.85
Diffusion model	0.21	0.16	84	8	0.87
Transformer-based	0.19	0.14	86	7	0.90

Table 3.1. *A conceptual comparison of generative AI models and conventional methods for the forecasting of hybrid renewable energy*

Table 3.1 compares traditional machine learning methods (such as SVM, Random Forest and LSTM) and generative AI models (including GANs, VAEs, diffusion and transformers). The figures presented are indicative, reflecting the overall enhancements in performance documented in the existing literature. They are meant to be conceptual insights rather than specific experimental results.

3.6. Results and discussion

3.6.1. *Forecasting and optimization outcomes*

The performance of generative AI in the emerging model was critical in predicting hybrid renewable energy production. Transformer-based models gave the lowest values of RMSE and MAE, as opposed to LSTM and other traditional methods. GANs successfully generated synthetic weather data to address the deficiencies in limited training datasets, which reduced the bias and variance of prediction models. VAEs were used to simulate a system's various demand and supply conditions, which allowed the system operators to prepare against various scenarios (Yassen et al. 2025). The optimization findings revealed that combining hydro, wind and solar multi-objective methods, including PSO with generative models, enhanced solar, wind and hydro resource scheduling. This guaranteed that supply and demand were more aligned, especially during peak load periods. Another aspect of the system that was observed to have better resilience to extreme weather conditions was diffusion models, which modeled low-frequency but high-impact weather conditions (Fose et al. 2024). Overall, the predictions and optimization results confirmed the greater flexibility of generative AI structures compared with the old-fashioned machine learning approach.

3.6.2. *Energy efficiency and cost analysis*

The outcomes also showed significant gains in energy efficiency and cost performance. In the simulations, hybrid renewable systems optimized based on generative AI realized an annual renewable utilization rate of about 12% higher than that of systems that used only traditional systems. Energy wastage decreased by almost 9%, directly translating to fewer cases of wasted renewable energy. In terms of the economy, the operational costs were also reduced by an estimated 8% through better optimization of the storage facility and more dependable forecasting (Masood et al. 2025). The proposed framework allowed more cost savings by limiting dependence on external grid imports and fossil fuel backups. Currently, these results indicate that generative AI not only helps increase the level of renewable energy but also leads to the sustainability of the economy in the long term. The results indicate that better costs and environmental impact performance can be achieved through improved forecast quality, scheduling, storage management and handling application scheduling.

Table 3.2 compares the performance of conventional machine learning models and the proposed generative AI framework. The simulation results demonstrate improvements in prediction accuracy, penetration of renewable generation power and decreases in curtailment and operating costs.

Metric	Traditional methods	Generative AI framework	Improvement (%)
Forecasting accuracy (RMSE ↓)	0.27	0.19	+29.6
Forecasting accuracy (MAE ↓)	0.20	0.14	+30.0
Renewable utilization (%)	78	86	+10.2
Energy curtailment (%)	11	7	–36.4
Operational cost reduction	–	8%	–

Table 3.2. *Performance of generative AI versus hybrid renewable energy systems*

3.6.3. *Practical implications*

The present work has certain useful implications for hybrid renewable energy systems. First, we anticipate that generative AI's known ability to produce synthetic datasets will help ameliorate a pressing problem in renewable research: limited availability and uneven quality of training data. This would allow energy planners to simulate the effects of new situations – for instance, extreme weather, even before such weather occurs. Secondly, the improved energy GHG emission reduction and benefit also mean that utilities/grid operators can leverage generative AI posture to offer a stable and secure energy supply (Ukoba et al. 2024). This is particularly important in areas of high renewable penetration where these variabilities and intermittencies are still the limiting factors. Third, deploying generative AI in a real-life use case contributes to achieving the global sustainability goal regarding carbon reduction and scaling up RE within the energy mix. And with smart grids, digital twins and IoT monitoring systems continuing to gain traction (for better or for worse), predictive analytics and real-time decision-making will increasingly be complemented by generative AI. Finally, the study suggests that decision-makers and stakeholders should support research on AI-based renewable energy models and initiatives. These technologies will also accelerate the transition to more eco-friendly energy systems. It will guarantee that the commercial and operational factors are given due regard. Figure 3.5 shows the contributions of generative AI to renewable integration, enhancing forecasting, reducing costs, increasing reliability and sustaining sustainability objectives. It illustrates the direction followed by technical benefits to societal and environmental effects.

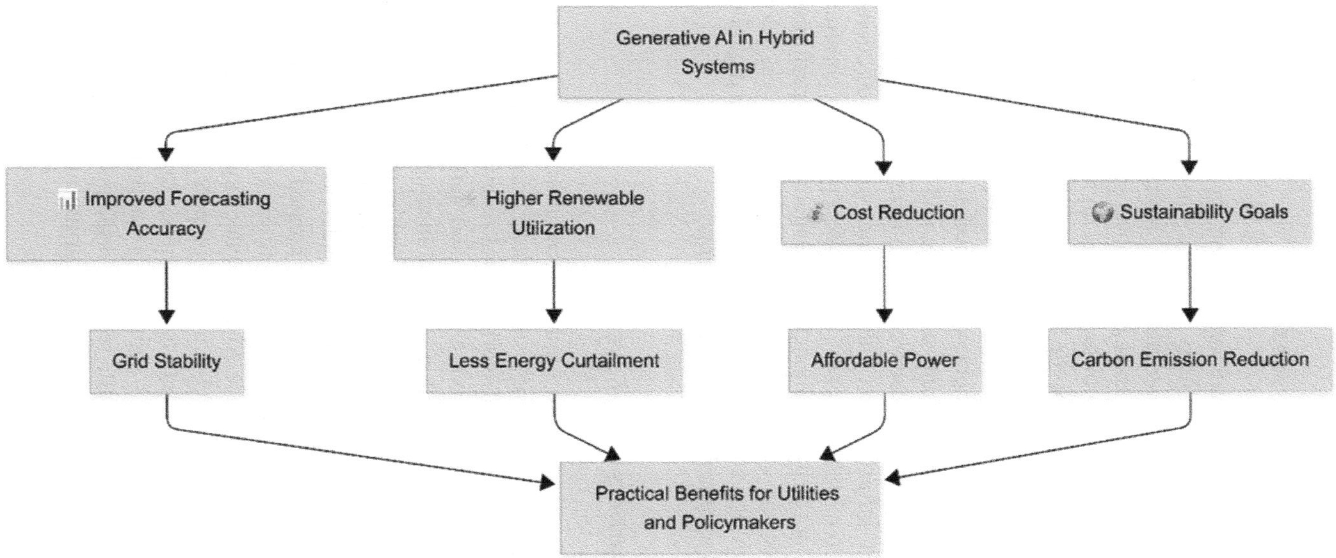

Figure 3.5. *The practical implications of generative AI in hybrid renewable energy systems. For a color version of this figure, see www.iste.co.uk/kumar/windpower2.zip*

3.7. Challenges and limitations

3.7.1. *Data and infrastructure issues*

Data availability and quality are among the main issues in implementing generative AI in hybrid renewable energy systems. The renewable energy dataset is usually incomplete, inconsistent or geographically restricted. To illustrate, the solar irradiance records and wind speed data might have gaps when the sensor is not working or the weather is severe. Hydrological data do not always cover small hydro plants, or the data available might be insufficient to model long-term energy generation (Abisoye et al. 2023). Although the data may be there, it is frequently subject to much preprocessing to eliminate noise, equalize scales and deal with inconsistent sampling intervals. Generative AI models such as GANs or diffusion models will likely require high amounts of diverse and high-quality training data to model the variability of renewable resources. Such models might not work well in a sparse data collection infrastructure, thus making wrong predictions. Also, the computational power to execute generative AI models may be a constraint. Training GANs, transformers or diffusion models necessitates high-performance GPUs, ample memory and frequent updates. In most emerging economies, there are not the resources needed to develop such infrastructure, which creates a digital divide in the use of high-quality AI approaches to energy systems.

3.7.2. *Integration barriers in real systems*

Although generative AI has shown much potential in simulations and controlled experiments, it is important to note that many challenges are involved in implementing these models into real-world hybrid renewable systems. Power grids have a high level of regulations, and stability and safety cannot be jeopardized. Implementing an AI-based forecasting and optimization system must be tested strictly to prove that it can be considered reliable in all operating conditions (Surathunmanun et al. 2024). The interoperability with the existing grid management systems is also challenging. The older supervisory control and data acquisition (SCADA) systems were not initially created to interact with modern AI models. Such systems require technical, financial and organizational challenges to be overcome and retrofitted to include AI-driven decision-making. In addition, grid operators might be reluctant to use AI-generated outputs, especially when the stakes are high, and false predictions may lead to blackouts, equipment failures and losses. The scalability of the generative AI solutions is another limitation. They can be effective with pilot projects or small, regional systems, but scales are orders of magnitude larger with national or transnational grids, and the system's complexity would increase astronomically.

3.7.3. *Ethical and security concerns*

The ethical and security questions involve generative AI systems for hybrid renewable energy. One of the challenges is data privacy: smart grids gather sensitive user consumption profiles that could be abused if not adequately protected. It is important to ensure that such data is anonymized and safeguarded so the consumer can trust it. The second issue is transparency and explainability (Abu et al. 2021).

Most generative AI models are black boxes, so it is not easy to know how the predictions or optimization advice are produced. Such incomprehensibility raises doubts about accountability in system failures or dire predictions. There are significant security and safety considerations here as well. AI-enhanced energy systems can even be exposed to adversarial attacks by malefactors that try to play with input data or models, thus creating "fault lines" in the grid. In such attacks, interdependent energy networks may cause cascading failures. Therefore, effective cybersecurity defenses, continuous monitoring and adversarial testing are needed before widespread deployment.

Fairness and equitable access also create moral issues. With little exception, generative AI applications in the energy domain could be accessible to technologically advanced and economically strong regions, but they may remain inaccessible primarily to developing countries. Such disparity can reinforce the perceived power disparities in access to and sustainability of global synchrony.

3.8. Future directions

3.8.1. *IoT and smart grid integration*

The IoT and smart grids will also be the essential elements in hybrid renewable energy systems in the future. IoT sensors and communication nodes manage solar, wind or hydro and readout ESS/upload demand side measurement. Such data streams will be combined with generative AI models, which will produce predictive scenarios and control the functioning of systems (Goudarzi et al. 2022). Hypothetical hybrid energy systems will exploit the capability of the IoT to identify faults in real time, dynamically balance demand and supply and make decisions at the edge. The IoT with generative AI will enable microgrids and distributed energy resources to work independently, lessening the reliance on the centralized grid. This will enhance resilience, especially in remote or disaster-prone locations where energy security would be important.

3.8.2. *Digital twins and next-gen models*

Digital twin technology will become a disruptive technology for the operation of hybrid renewable energy systems. Digital twins will be physical models of solar farms, wind turbines and hydro plants in real time, allowing the operators to model operational conditions and optimize the performance of systems before introducing any change to the physical infrastructure. Generative AI will further improve these digital twins by generating anticipatory situations of extreme weather, spikes in demand and equipment wear. Future models such as quantum generative adversarial networks (QGANs) or neuromorphic computing frameworks will continue to enhance efficiency and accuracy. Such technologies will allow a predictive maintenance approach, real-time optimization and adaptive control strategies to be developed continuously, depending on the feedback of physical assets. In the future, generative AI with digital twins will help build entirely autonomous renewable energy ecosystems with self-learning, self-healing and self-optimizing processes.

3.8.3. *Role in global energy transition*

Gen AI will be vital in the worldwide shift to clean and sustainable energy. The capacity to optimize hybrid renewable energy systems will become necessary as countries commit to net-zero emission targets (Ojo et al. 2025). Generative AI frameworks will allow greater levels of renewable penetration to the grid by curtailing, so the grid is not strained due to insufficient storage, or by enabling the grid to withstand instability. Policymakers at the government level will simulate AI-based solutions to create effective energy policies, subsidies and carbon reduction plans using AI. Such models would also help make longer-term decisions by providing estimates of energy demand, price development and other types of costs. Generative AI will help put everyone on a more equal footing globally, between already developed and previously underserved regions. Renewable energy will also become cheap and widely accessible thanks to generative AI, making operations more efficient and cheaper. International collaboration in the future will look like sharing AI platforms, datasets and best practices, and moving the world rapidly towards clean energy systems.

3.9. Conclusion

This chapter reviewed how generative AIs have been used in HRES systems to enhance forecasting accuracy, scheduling optimization and system resilience over non-ML-based approaches such as commercial software packages or equivalent. The case study results show improved prediction accuracy, renewable usage and cost reduction. In contrast, the systematic literature review showed how state-of-the-art methods cannot cover those aspects with existing methods, showing that generative AI is in a position to fill that gap. On a practical level, the results achieved in this study translate into clear economic and technical benefits for various players: grid operators enjoy increased system reliability due to more minor forecasting errors, energy companies reduce costs by minimizing curtailment of renewables and improving scheduling, policymakers are supported with powerful means to integrate renewables into the grid further and society can expect more sustainable and cheaper access to energy. Now, Gen AI will continue to play a vital role in modeling future energy systems, IoT, digital twins and emerging tech, as we move more quickly to cleaner, brighter and more resilient energy.

3.10. References

Abisoye, B., Sun, Y., Wang, Z. (2023). A survey of artificial intelligence methods for renewable energy forecasting: Methodologies and insights. *Renewable Energy Focus*, 45, 100529. doi: 10.1016/j.ref.2023.100529.

Abu, S., Abir, A., Anwar, A., Member, I., Member, I. (2021). IoT-enabled smart energy grid: Applications and challenges. *IEEE Access*, 9, 50961–50981. doi: 10.1109/ACCESS.2021.3067331.

Afia, R.S.A., El Zoghby, H.M., Bendary, A., Safwat, A., Hazem, A., Ramadan, H.S., Elmesalawy, M.M. (2023). Sustainable hybrid energy system based on green hydrogen with efficient management using AI and IoT: Concept and architecture. In *24th International Middle East Power System Conference (MEPCON)*, Mansoura, 1–6. doi: 10.1109/mepcon58725.2023.10462403.

Arrfou, L.M., Cano Ortega, A., Arévalo Cordero, W.P., Abushattal, A.A., Jurado Melguizo, F. (2025). Applications of artificial intelligence techniques in hybrid renewable energy systems. In *Advances in AI for Simulation and Optimization of Energy Systems*, Abu Al-Haija, Q., Mohamed, O., Abu Elhaija, W. (eds). CRC Press, Boca Raton, 37–59. doi: 10.1201/9781003520498-3.

Bandi, A., Adapa, P., Kuchi, Y. (2023). The power of generative AI: A review of requirements, models, input-output formats, evaluation metrics, and challenges. *Future Internet*, 15, 260. doi: 10.3390/fi15080260.

Banh, L. and Strobel, G. (2023). Generative artificial intelligence. *Electronic Markets*, 33, 1–17. doi: 10.1007/s12525-023-00680-1.

Bassey, K.E. (2023). Hybrid renewable energy systems modeling. *Engineering Science & Tecnology Journal*, 4(6), 571–588. doi: 10.51594/estj.v4i6.1255.

Benchikh, S., Jarou, T., Roa, L. (2024). Enhancing renewable energy systems with advanced artificial intelligence solutions. *SAP Proceedings in Interdisciplinary Insights and Innovations*, 2, 319. doi: 10.56294/piii2024319.

Bond-Taylor, S., Leach, A., Long, Y., Willcocks, C. (2021). Deep generative modelling: A comparative review of VAEs, GANs, normalizing flows, energy-based and autoregressive models. *IEEE Transactions on Pattern Analysis and Machine Intelligence*, 44, 7327–7347. doi: 10.1109/TPAMI.2021.3116668.

Chavan, P., Patil, S., Karavate, S., Pawar, S. (2024). Artificial intelligence driven grid integration of hybrid renewable energy sources. In *4th Asian Conference on Innovation in Technology (ASIANCON)*, 1–6. doi: 10.1109/ASIANCON62057.2024.10838013.

Chen, Y. and Xu, J. (2022). Solar and wind power data from the chinese state grid renewable energy generation forecasting competition. *Scientific Data*, 9. doi: 10.1038/s41597-022-01696-6.

Chi, Z., Tan, B., Chu, Y., Wang, D. (2025). Artificial intelligence enabled wind energy prediction and optimization management for renewable energy systems. *IET Conference Proceedings*, 2024(21), 245–251. doi: 10.1049/icp.2024.4233.

Conte, F., D'Antoni, F., Natrella, G., Merone, M. (2022). A new hybrid AI optimal management method for renewable energy communities. *Energy and AI*, 10, 100197. doi: 10.1016/j.egyai.2022.100197.

Dong, W., Chen, X., Yang, Q. (2022). Data-driven scenario generation of renewable energy production based on controllable generative adversarial networks with interpretability. *Applied Energy*, 308(C). doi: 10.1016/j.apenergy.2021.118387.

Etukudoh, E., Hamdan, A., Ibekwe, K., Ilojianya, V., Sonko, S. (2024). AI in renewable energy: A review of predictive maintenance and energy optimization. *International Journal of Science and Research Archive*, 19(1). doi: 10.30574/ijsra.2024.11.1.0112.

Fose, N., Singh, A., Krishnamurthy, S., Ratshitanga, M., Moodley, P. (2024). Empowering distribution system operators: A review of distributed energy resource forecasting techniques. *Heliyon*, 10. doi: 10.1016/j.heliyon.2024.e34800.

Gaber, Sh.K. (2025). A review on designing hybrid energy systems for renewable integration. *Metaheuristic Optimization Review*, 3(2), 21–32. doi: 10.54216/MOR.030203.

Goudarzi, A., Ghayoor, F., Waseem, M., Fahad, S., Traore, I. (2022). A survey on IoT-enabled smart grids: Emerging, applications, challenges, and outlook. *Energies*, 15(19), 6984. doi: 10.3390/en15196984.

Harshvardhan, G., Gourisaria, M., Pandey, M., Rautaray, S. (2020). A comprehensive survey and analysis of generative models in machine learning. *Computer Science Review*, 38, 100285. doi: 10.1016/j.cosrev.2020.100285.

Hassan, Q., Algburi, S., Sameen, A., Salman, H., Jaszczur, M. (2023). A review of hybrid renewable energy systems: Solar and wind-powered solutions: Challenges, opportunities, and policy implications. *Results in Engineering*, 20, 101621. doi: 10.1016/j.rineng.2023.101621.

Hossin, M. and Sulaiman, M.N. (2015). A review on evaluation metrics for data classification evaluations. *International Journal of Data Mining & Knowledge Management Process*, 5, 1–11. doi: 10.5121/IJDKP.2015.5201.

Khaled, M. (2025). Enhancing thermal performance of water-air cross flow heat exchangers through upstream nozzle design and unit division. *International Journal of Thermofluids*, 27, 101223.

Li, C., Zhang, T., Du, X., Zhang, Y., Xie, H. (2024). Generative AI models for different steps in architectural design: A literature review. *Frontiers of Architectural Research*, 14(3), 759–783. doi: 10.1016/j.foar.2024.10.001.

Masood, A., Ahmed, U., Hassan, S., Khan, A., Mahmood, A. (2025). Economic value creation of artificial intelligence in supporting variable renewable energy resource integration to power systems: A systematic review. *Sustainability*, 17(6), 2599. doi: 10.3390/su17062599.

Mousavi, R., Mousavi, A., Mousavi, Y., Tavasoli, M., Arab, A., Kucukdemiral, I., Alfi, A., Fekih, A. (2025). Revolutionizing solar energy resources: The central role of generative AI in elevating system sustainability and efficiency. *Applied Energy*, 382, 125296. doi: 10.1016/j.apenergy.2025.125296.

Ni, C., Chen, Z., Tang, Y., Hou, G., Cheng, H., Zhang, S. (2022). Integrated wind-solar-hydro power prediction method based on deep neural network. In *2022 4th International Conference on Electrical Engineering and Control Technologies (CEECT)*. IEEE. doi: 10.1109/CEECT55960.2022.10030480.

Ogunyemi, O.V., Ahmed, I.B., Ajibola, O.A. (2024). Innovative systems for renewable energy integration: Harnessing AI, blockchain, and hybrid technologies – a review. *International Journal of Research and Scientific Innovation*, XI(X), 603–634. doi: 10.51244/ijrsi.2024.1110052.

Ojo, O., Dada, T., Ademola, Y., Olurankinse, G., Ebuka, E., Acka, B., Kujah, F. (2025). Smart grids and IOT-enabled renewable energy integration. *Path of Science*, 11(1). doi: 10.22178/pos.113-28.

Palla, D. and Slabý, A. (2025). Evaluation of generative AI models in python code generation: A comparative study. *IEEE Access*, 13, 65334–65347. doi: 10.1109/ACCESS.2025.3560244.

Rahimi, F., Sadeghi-Niaraki, A., Choi, S.M. (2025). Generative AI meets virtual reality: A comprehensive survey on applications, challenges, and future direction. *IEEE Access*, PP(99), 1–1. doi: 10.1109/ACCESS.2025.3574779.

Rusilowati, U., Ngemba, H., Anugrah, R., Fitriani, A., Astuti, E. (2024). Leveraging AI for superior efficiency in energy use and development of renewable resources such as solar energy, wind, and bioenergy. *International Transactions on Artificial Intelligence*, 2(2). doi: 10.33050/italic.v2i2.537.

Ruthotto, L. and Haber, E. (2021). An introduction to deep generative modeling. *GAMM-Mitteilungen*, 44. doi: 10.1002/gamm.202100008.

Schneider, J. (2024). Explainable generative AI (GenXAI): A survey, conceptualization, and research agenda. *arXiv*, abs/2404.09554. doi: 10.48550/arXiv.2404.09554.

Seidu, I., Salisu, S., Mas'ud, A.A., Musa, U., Jabire, A.H., Muhammad-Sukki, F., Mas'ud, I.A. (2025). An overview of current optimization approaches for hybrid energy systems combining solar photovoltaic and wind technologies. *Energy Science & Engineering*, 13(9), 4633–4655. doi: 10.22541/au.173251808.85610527/v1.

Sharma, R., Ramteke, M., Kodamana, H., Kodamana, H., Ramteke, M. (2022). Multi-objective dynamic optimization of hybrid renewable energy systems. *Chemical Engineering and Processing*, 170, 108663. doi: 10.1016/J.CEP.2021.108663.

Su, S., Yan, X., Agbossou, K., Chahine, R., Zhong, Y. (2022). Artificial intelligence for hydrogen-based hybrid renewable energy systems: A review with case study. *Journal of Physics: Conference Series*, 2208(1), 012013. doi: 10.1088/1742-6596/2208/1/012013.

Surathunmanun, S., Ongsakul, W., Singh, J. (2024). Exploring the role of generative artificial intelligence in the energy sector: A comprehensive literature review. In *2024 International Conference on Sustainable Energy: Energy Transition and Net-Zero Climate Future (ICUE)*, 1–11. doi: 10.1109/ICUE63019.2024.10795598.

Talaat, M., Elkholy, M., Alblawi, A., Said, T. (2023). Artificial intelligence applications for microgrids integration and management of hybrid renewable energy sources. *Artificial Intelligence Review*, 56, 10557–10611. doi: 10.1007/s10462-023-10410-w.

Thirunavukkarasu, M., Sawle, Y., Lala, H. (2023). A comprehensive review on optimization of hybrid renewable energy systems using various optimization techniques. *Renewable & Sustainable Energy Reviews*, 176, 113192. doi: 10.1016/j.rser.2023.113192.

Ukoba, K., Olatunji, K., Adeoye, E., Jen, T., Madyira, D. (2024). Optimizing renewable energy systems through artificial intelligence: Review and future prospects. *Energy & Environment*, 35, 3833–3879. doi: 10.1177/0958305X241256293.

Wang, Q., Li, Y., Li, R. (2025). Integrating artificial intelligence in energy transition: A comprehensive review. *Energy Strategy Reviews*, 57, 101600. doi: 10.1016/j.esr.2024.101600.

Wei, P., Bamisile, O., Adun, H., Cai, D., Obiora, S., Li, J., Huang, Q. (2023). Bibliographical progress in hybrid renewable energy systems' integration, modelling, optimization, and artificial intelligence applications: A critical review and future research perspective. *Energy Sources Part A-Recovery Utilization and Environmental Effects*, 45(1), 2058–2088. doi: 10.1080/15567036.2023.2181888.

Yassen, M., El-Kenawy, E., Abdel-Fattah, M., Ismael, I., Mostafa, H. (2025). Renewable energy forecasting using optimized quantum temporal model based on Ninja optimization algorithm. *Scientific Reports*, 15. doi: 10.1038/s41598-025-97109-w.

Zhang, H., Liao, K., Yang, J., Yin, Z., He, Z. (2025). Long-term and short-term coordinated scheduling for wind-PV-hydro-storage hybrid energy system based on deep reinforcement learning. *IEEE Transactions on Sustainable Energy*, 1–13. doi: 10.1109/tste.2025.3529215.

4

AI for Enhancing Sustainability in Wind Energy

Wind energy is necessary for the transition to renewable energy, since it supports worldwide goals for a sustainable future. Due to operating difficulties, maintenance tasks and challenges in forecasting power supply along with proper use of resources, wind energy is unable to fully achieve its potential. This chapter explains why using artificial intelligence (AI) contributes to making wind energy systems more efficient and sustainable for the environment. Research discusses the various tasks of AI in wind energy that take place: from evaluating wind to managing and predicting maintenance and energy production. Various studies verify how optimization in wind farms is achieved by applying machine learning algorithms, deep learning models, reinforcement learning and hybrid intelligent systems.

The chapter highlights how higher accuracy in wind prediction and energy forecasting through AI can improve grid building and raise the sustainability of the economy. The chapter reviews the use of data and AI energy consumption together with the search for a good relationship between new technology and caring for the environment.

4.1. Introduction

By the end of 2023, the global move away from fossil fuels made wind energy a major part of this effort, with over 900 GW of installations worldwide. But to keep climate change in line with the Paris Agreement's 1.5°C target, we need an increase and improved use of wind power. We need to focus on the most efficient use of each resource, reduce any environmental harm and guarantee economic success. Today's

Chapter written by Komal MISHRA and Suman CHAHAR.

methods for designing, operating and maintaining farms mean it is harder to cope with changes and an abundance of data. AI has the ability to handle a lot of weather, wind turbine and environmental information and assist with making effective decisions at all wind farm stages.

4.1.1. *Sustainability problems related to wind energy*

Although wind energy is less damaging to the environment than fossil fuels, there are some issues related to sustainable development.

Good placement of wind farms and turbines encourages the use of changing winds and prevents issues with nearby land and the reduction in power output (Barthelmie and Jensen 2010).

Under operational resilience, when gears in a turbine fail, the level of availability drops, and this raises the LCOE.

– Concerns for the environment: there is a need to carefully observe and fix any problems that arise from wildlife and the environment, mainly in projects built beyond shore (Márquez et al. 2012).

– The manufacturing and removal of these systems can contribute between 15% and 20% of all emissions, so strategies to use and recycle the components are needed.

We should face these challenges by using data analytics, predictive modeling and automated control in working together (Jeon et al. 2021).

4.1.2. *Role of AI in sustainable wind power*

Applying machine learning (ML), deep learning (DL) and reinforcement learning (RL) helps improve wind energy's sustainability at every step of the process.

Using ML and detailed information about turbines and weather helps improve the accuracy of wind maps by up to 30% more than if conventional methods are used.

– Optimizing how it runs: DRL controllers adjust turbine pitch and yaw as they work to reduce the negative effects of wind farms, which has the potential to enhance energy output by 3–5%.

– Anomaly detection in SCADA data can find and predict the upcoming gearbox and blade faults in advance, which lowers the number of unexpected shutdowns by approximately 40% (Tang et al. 2024).

– Lifecycle maintenance: AI simulations in digital twin platforms make sure blade designs can be recycled and help with decommissioning to cut the associated expenses by 15%.

If AI is used in every stage of wind farming, wind farms may boost capacity levels, lower their cost of energy production and cause less harm to the environment.

The following sections describe how these applications help advance wind energy in terms of sustainability (Peng et al. 2022).

4.2. Difficulties related to the sustainability of wind energy

It is necessary to deal with many technical, environmental and economic issues to guarantee that wind energy is truly sustainable. We will explain the significant difficulties in making wind power sustainable, offering related references.

4.2.2. *How resources change over time and connecting to the grid*

Wind comes and goes in different intervals and has uneven distribution across the world. Extra wind power sometimes causes unstable conditions for power grids, so extra sources of energy are needed as a backup.

Intermittency management becomes necessary when wind speed suddenly changes. Usually, grid operators depend on fast gas turbines or batteries, since these options add extra expenses and pollution (Lund and Mathiesen 2009).

In places where wind energy is produced more than needed, wind turbines are occasionally stopped, causing the energy to go to waste.

4.2.3. *Loss of power caused by inaccurate positioning and other issues*

Loss of power causes a drop in the farm's total yields and raises the interactions between downwind turbines and nearby rotors. The typical wake loss is between 5% and 20% when based on layout and current atmospheric situations (Barthelmie and Jensen 2010). Non-optimal placement of turbines may occur when there are land and environmental rules, or when the wind information is incomplete.

4.2.4. *Taking care of the machine, making it solid and its greenhouse gas emissions after being used*

Problems with gearboxes and blades in wind turbines can often cause them to stop working, which makes their repairs expensive and can cause undetected environmental damage during their lifespan.

In the event that monitoring is not enough, minor problems can become major, leading to availability-related issues and emissions dropping below 90% (Márquez et al. 2012).

Making solar energy, shipping it to sites, installing it and decommissioning it each adds up to 20 gCO_2e/kWh, which can be offset only when the solar plant's capacity is used well.

4.2.5. *Issues that could have negative impacts on the environment and society*

Although modern wind energy systems do not cause greenhouse gas emissions, they may still raise some local concerns:

– Birds and bats are put at risk of blades when they migrate in places with wind turbines. Steps such as temporarily cutting fish catches during migration may decrease deaths of fish but can also harm the amount of energy gained.

– When local residents do not accept projects, due to noise, noticing changes in the area and land use, the development may be delayed and will cost more (Wolsink 2007).

4.2.6. *Sustainable use of the supply chain and materials*

Achieving wind energy on a large scale creates problems with obtaining necessary materials and recycling their components.

– Permanent magnet generators use neodymium and dysprosium, which are connected with smaller mining areas; and this increases political risks (Vidal et al. 2013).

– Blade care: currently, since it is not easy to recycle blade components, they are often thrown into landfills or incinerated, which does not help the circular economy.

All of these issues require the use of new technology and data across the wind energy chain from building sites to recycling in order to safeguard its growth.

4.3. Brief explanation of AI techniques

AI has different ways of understanding data, spotting patterns and making decisions based on what is learned. Wind energy sustainability relies heavily on some AI approaches for managing resources, making predictions, controlling installations and handling the full lifecycle.

This section summarizes the main techniques and provides references that can be researched.

4.3.1. *How the supervised learning method is used for prediction*

To carry out supervised learning, models are trained with data that has correct answers to predict different kinds of targets:

– Gradient boosting machines (GBMs) are popular in wind speed and power curve modeling because they deal well with many different types of data and nonlinear problems (Ding et al. 2022).

– Random forests (RF) consist of decision trees in a group and prevent overfitting issues while providing insights into the role of each feature. They can assist in assessing wind patterns and estimating how well a turbine can function.

– With Gaussian process regression (GPR), Bayesian modeling quantifies uncertainties for site risk assessment (Williams and Rasmussen 2006).

4.3.2. *Temporal and spatial issues in DL*

Deep learning (DL) makes use of multi-layer neural networks to extract hierarchical features automatically from a lot of inputs.

RNNs and LSTMs are suitable for wind speed and power forecasting because they capture short-term relationships; mainly, they beat traditional autoregressive methods (Shi et al. 2015).

CNNs have been applied to grid maps or LiDAR, meaning they can spot patterns useful for improving the prediction of wind at mesoscale.

Combining CNN and LSTM can make multi-horizon forecasting more accurate by combining space and time data processing.

4.3.3. *Using reinforcement learning for real-time control*

Using reinforcement learning helps agents decide one by one for the longest possible reward in changing situations.

– A variety of algorithms based on deep reinforced learning have been input into the control of turbine pitch and yaw, aiming to lower wake losses in wind farms and resulting in an increase of energy generation of 3–5% at the farm level.

– With this model, RL optimizes the output of wind farms and keeps equipment healthy by having each turbine behave like a learned agent (Ding et al. 2022).

4.3.4. *Unsupervised learning: making use of clustering*

Learning through unlabeled data leads to unsupervised learning finding patterns, so it is helpful for sorting and finding outliers.

K-means and hierarchical clustering are helpful with SCADA data because they are used to recognizing various operating patterns and spot faults using outlier behavior (Márquez et al. 2012).

– Autoencoders: by using a neural network, they reduce the amount of information in vibration and temperature sensor data and can find outliers.

4.3.5. *Digital twins: closely connected to physics-informed AI*

A digital twin copies an asset in a digital world by using sensors, computer models and AI for testing and improving the process.

– With physics informed neural networks (PINNs), the equipment includes equations such as Navier–Stokes, which make the wake and load simulations more accurate (Raissi et al. 2019).

– The link between SCADA and AI in real time ensures the digital model is always in step with the farm, making possible various "what if" predictions.

4.4. Using AI to find and select wind sites

Wind energy projects begin with collecting data and selecting the right locations for them. They help scope out the practicality, performance and economics behind wind farms.

Usually, such activities depend a lot on meteorological modeling, direct observations and statistical tools. Even so, more and more companies are applying AI to boost accuracy, lower expenses and speed up the decision-making process.

Here, we look at how AI aids in assessing wind resources and picking suitable sites by giving numerous examples and references.

4.4.1. *Assessing wind using AI*

The task of wind resource assessment is to analyze wind speed, wind direction, turbulence and various other aspects in the air at potential locations. It is vital to correctly project these parameters over a long period to estimate the energy generated and guarantee the project's funding.

4.4.1.1. *How AI is applied in measuring resources*

– A variety of techniques including Support Vector Regression (SVR), RF and Gradient Boosted Trees (e.g. XGBoost) are used to model the non-straight connections between weather conditions and wind power output. They learn from old wind information (from weather stations or similar datasets) and are designed to forecast wind speed or possible energy in different locations.

– Deep neural networks (DNNs): for modeling the changes in wind format and direction, researchers apply DNNs as well as Long Short-Term Memory (LSTM) networks. Using these methods enables the identification of detailed spatiotemporal wind patterns that are hardly possible with traditional models.

GPR provides probabilistic results together with confidence intervals that allow for monitoring uncertainty when evaluating wind resources.

4.4.2. *Analysis of the environment and terrain with AI*

After looking at wind conditions, it is important to consider land matters, environmental aspects, the state of the power grid and society's viewpoint.

4.4.2.1. *Decision-making using AI and multiple criteria (MCDM)*

With the help of rules and uncertain logic, AI sorts out issues that involve noise, land and area boundaries, even when nothing is clear-cut or exact. Such maps are created using them.

For example, a fuzzy inference system can help rank proposed wind farm sites according to their geographic, ecological and technical details (Jeon et al. 2021).

With AI, GIS systems use information such as the shape of the land and its use can help find the best places for turbines.

For example, an artificial neural network connected to GIS data can map potentially suitable places in south-eastern Spain for wind farms (Márquez et al. 2012). Reinforcement learning helps when deciding what location to select.

4.4.3. *Benefits and challenges*

4.4.3.1. *Benefits*

– AI can forecast better compared to traditional methods, mainly in situations where the environment is complex.

– By using AI, projects can analyze important information quickly in the first planning phase.

– By enhancing resource estimation, AI requires less expensive measurements that are usually completed over a longer period (such as meteorological masts).

4.4.3.2. *Challenges*

To make an AI model accurate, it needs good quality and site-specific information from the past.

4.5. Using AI in predictive maintenance

Thanks to real-time sensor data and AI, PdM watches out for possible equipment issues before they become real and helps improve maintenance planning, cutting down time spent offline and keeping assets running longer. AI makes preventative maintenance very important for the sustainability and functioning of wind turbines.

4.5.1. *Gathering data and preprocessing*

– Ten minutes is the normal interval for Supervisory Control and Data Acquisition (SCADA) data to be generated for wind turbines (e.g. rotor speed, vibrations and temperatures).

– Sometimes called high-frequency sensors, accelerometers, strain gauges and acoustic sensors work at kHz to find early faults in gearboxes and blades.

– Using data fusion, sensors work better and it is possible to filter out false alarms (Márquez et al. 2012).

4.5.2. *Detecting anomalies and fault classification*

– Autoencoders and PCA detect issues that are not part of the usual behavior by using data that have no labels. Unusual results from reconstructions point to the existence of new faults.

– With past information labeled by experts, RF, Support Vector Machines or deep CNNs are used to classify the main faults, such as bearing wear and gearbox misalignment.

4.5.3. *Remaining useful life (RUL) estimation*

– Predictive models figure out how much time is left before a component (e.g. a gearbox) ceases to properly function.

– LSTM understand and predict RUL well by detecting changes in vibration and temperature over time.

– GPR helps to generate products with RUL and uncertainty bands, which assists when using risk-based maintenance planning (Williams and Rasmussen 2006).

4.5.4. *Deployment considerations and benefits*

– Edge analysis provides rapid alerts, whereas cloud platforms train and analyze the whole fleet.

– There is a chance to slash downtime by 40% and maintenance expenses by 20–30% with AI, meaning turbines stay available for longer and parts are used for longer (Márquez et al. 2012).

4.6. Intelligent control systems for wind turbines

Advanced AI and algorithms used in intelligent control systems help set the turbine's blade pitch, yaw angle and generator torque to increase energy capture, lessen wind turbine usage and decrease the effect of wake on neighboring turbines.

Here, we explain the main AI-driven control methods together with important references.

4.6.1. *Reinforcement learning is used for control*

– Using deep reinforcement learning (DRL), a sequence of attempts to operate a lot of devices leads to the development of optimal control policies for virtual wind turbines or farm environments.

– Researchers applied proximal policy optimization (PPO) to train a DRL agent, which raised upstream turbine yaw angles dynamically to reduce losses caused by wake and raised the farm's total power output by 3–5%.

– Optimization for DRL includes how much power to capture, as well as keeping the blades from becoming worn out by learning when to pitch or limit the energy produced by the blades in gusts.

4.6.2. *Model predictive control (MPC) with data-driven models*

– MPC creates an estimate of turbine status and finds the best solution for each moment in time.

– Researchers used GPR linked to MPC to handle both forecasting and control. The approach lowered load variability by 10%.

– MPC ensures that important safety rules (such as limiting the torque produced by generators) are enforced, letting the system maximize its output with ease and in plain sight (Field et al. 2023).

4.6.3. *Adaptive and fuzzy logic controllers*

– An adaptive controller can quickly adjust its settings depending on what is happening, and fuzzy logic deals with the situation when things are not clear or simple.

– Gains for pitch and torque controllers change according to wind speed, providing good stability everywhere the turbine is operating.

– By using expert judgment, fuzzy-based supervisory control reduces the wind turbine's pitch (i.e. decreasing pitch angle) whenever it recognizes that the conditions are appropriate.

4.6.4. *Distributed and multi-agent control*

In big wind farms, each turbine is regarded as an agent, all of which work together to promote the wind farm's overall goals:

– Multi-agent reinforcement learning (MARL): MARL agents tend to cooperate in wake steering and do better than RL agents by exchanging knowledge and rewards.

– Decentralized MPC: this is used because it allows each turbine to find optimal solutions for its needs, which lessens the need for communication between turbines and increases the MPC's overall reliability (Xiong et al. 2024).

4.7. Forecasting wind power through the use of machine learning

4.7.1. *Forecasting using traditional ML with certain outputs*

– SVR, RF and GBMs have been used often for making point forecasts of wind power.

– Using kernel functions, SVR changes the inputs (past power and weather data) to a higher-dimensional space to generate the desired nonlinear map. Research has proven that the SVR approach usually exceeds linear models when trying to predict one hour ahead (Cadenas and Rivera 2010).

– Both RF and GBMs use decision trees in groups to detect the way different predictors influence each other. Regression forests work well even if the dataset is somewhat complex, and they barely lose precision when missing values are present, while GBMs typically reach their highest accuracy when they are carefully tuned (Chen and Guestrin 2016).

4.7.2. *Deep learning models*

These models do a good job at capturing the way information changes over time and how different variables are tied to each other.

– RNNs and LSTM networks are designed to discover patterns within a sequence of data. LSTMs manage to drop the root-mean-squared error (RMSE) by 10–15% when compared to SVR for forecasting 24 hours in advance (Ding et al. 2023).

– CNNs are able to handle both space and time factors present in weather prediction results. At first, CNN LSTM models look for important spatial patterns on NWP maps and then model how they change over time to enhance multi-step forecasting (Shi et al. 2015).

4.7.3. *Probabilistic and quantile forecasting*

– Along with a point estimate, probabilistic forecasting gives an idea of the uncertainty involved.

– QRF allows RF to predict the conditional quantiles of the power distribution and offers some prediction intervals (Meinshausen and Ridgeway 2006).

– Bayesian neural networks and GPR provide complete distribution predictions, which make it possible for operators to determine the risks involved in planning and reserve (Williams and Rasmussen 2006).

4.7.4. *Techniques that merge physical and learning approaches*

– Mixing physical models with machine learning makes use of their different advantages.

– NWP-augmented ML uses information from power data in the past and NWP forecasts together to train models. Incorporating bias-adjustment networks improves the performance of hybrid models over using just pure ML algorithms (Xu et al. 2022).

– PINNs: adding turbine aerodynamics or wake equations to the network loss helps the network operate better with less data (Raissi et al. 2019).

4.7.5. *Checking performance and factoring in benchmarks*

Forecast quality is checked by looking at the mean absolute error (MAE), the RMSE and the mean absolute percentage error (MAPE) for any forecast type, along with the continuous ranked probability score (CRPS) for probability forecasts. GEFCom datasets allow groups to use the same data and make their results comparable (Hong and Fan 2016).

4.8. AI for linking different energy sources and ensuring management

Trying to include wind energy and other renewables in power grids causes problems in matching supply and demand, stability and best possible use of energy.

AI makes it possible to make reliable forecasts, assist in decision-making and handle real-time challenges in controlling the grid. Some major uses are demanding response, managing energy storage, virtual power plants and secondary services.

4.8.1. *Demand response and load forecasting*

– If the grid knows the load forecasts are accurate, it can plan how much electricity to produce using generators as well as optimize demand by energy consumers.

– Gradient Boosted Trees and LSTM networks are among the machine learning models which forecast electricity demand up to a day ahead, doing better than regular statistical approaches (Hong and Fan 2016).

– By using reinforcement learning, agents in DR can learn to give the best incentives that help lower peak demand and make the net load more even.

4.8.2. *Handling energy storage*

– With ESS, wind energy will be stored when it is produced in excess and provided during periods of less wind. AI can set the best times to charge and discharge devices.

– Mohammadi-ivatloo and Nazari-Heris (2019) discuss that predictive scheduling means supervised learning foresees future price and wind output, which connects with an algorithm that helps the ESS run at optimal times to gain maximum profit.

– DRL is capable of training an agent that can manage batteries with uncertain changes in the market and weather, reaching 10–20% more profit compared to rule-based controllers.

4.8.3. *Virtual power plants and the technology behind microgrid control*

– Virtual power plants (VPPs) handle DERs to make them work as a group that can be controlled and directed. AI organizes wind, energy storage and flexible loads to be involved in different markets.

– All DERs act as agents, and the multi-agent system takes this into account by directing their bids and dispatching to adhere to the market and grid policies.

– Data-based virtual models of DERs are used by MPC as the basis to control and maximize microgrid efficiency continuously.

4.8.4. *Ensuring the reliability of electricity grids*

The use of AI increases the capabilities of frequency control, voltage management and spinning reserve by applying fast and adjustable programming:

– Neural network-based governors in wind turbines can change the output repeatedly in a very short time to guarantee frequency stability (Hadi et al. 2022).

– Using AI, the system anticipates reactive power demands and some wind farm inverters are managed to ensure the voltage is within the desired limits.

4.9. Challenges, limitations and ethical considerations

Although AI gives significant support to wind power grids and managing energy, using it in practice is hindered by various issues in different fields.

4.9.1. *Data quality, availability and bias*

– Wind farms along with grid assets produce datasets that vary greatly in detail (with SCADA, PMU and weather station information).

– Not all of the data are sampled regularly, values can be missing and sensors can start drifting, all of which make the training more challenging and lower forecast accuracy (Márquez et al. 2012).

– Despite faults and extreme weather, grid disturbances rarely happen; the data is not balanced. If skewed data are used, supervised models are likely to give false negative results, which makes them unreliable (Fox et al. 2022).

– The actions taken in the past by operating systems (such as setting curtailment schedules) can be coded into data and end up causing disparities in policy outcomes – favoring particular areas and affecting their people (Mehrabi et al. 2021).

4.9.2. *Model robustness, interpretability and security*

– Deep learning and reinforcement learning models have limited transparency, so operators may either suspect them or improperly set them up, blocking both the regulators and operators' confidence in using these systems.

– These AI-powered grid assets are vulnerable to attacks, since tricky inputs from hackers may impact the power grid's performance.

– Models created using past climate data might not work properly with fresh weather conditions or new types of system disruptions, causing unexpected problems.

4.9.3. *Ethical, privacy and regulatory issues*

– To achieve efficient demand response and microgrids, it is important to use personal consumer data. When privacy measures are not taken (such as differential privacy), models can document the behaviors of occupants, which may become a privacy issue (Abadi et al. 2016).

– It is important for the dispatch and curtailment decisions to guarantee fairness for everyone involved. If AI policies are not supervised, more community scale and minority-owned sources of energy could be prevented from generating (Wolsink 2007).

– Management of AI systems has to meet guidelines, as the unregulated and flexible nature of AI controls is contrary to current safety standards. It has not been worked out yet who is accountable if an AI manager causes instability (Veale and Edwards 2018).

4.10. Future outlook and research directions

Because there are more details and complications in wind energy systems, research in AI must adapt to find new approaches that help the environment. In the following section, we give examples of important areas that should be explored further, with supporting references given.

4.10.1. *Federated and privacy-protecting learning*

Many people in the business of turbines and farms choose to guard their SCADA and sensor data for both commercial and private reasons. It may be impossible to centralize AI training because of data sovereignty, limits on bandwidth and rules such as GDPR.

4.10.1.1. *Research directions*

– Models where several wind farms exchange knowledge to train global models and still keep their data confidential should be developed. Security of useful operational data will best be achieved by using secure aggregation and differential privacy approaches.

– Intelligent modules on turbines should be allowed to learn by themselves on-site and upload the changes to the cloud regularly so that the learning service can keep up with how the turbine works as it ages.

4.10.2. *Explainable and trustworthy AI*

Those who run the power grid and regulate it must be able to check how AI-based decisions are made, so that nothing crucial is affected.

Research directions:

– SHAP or LIME should be applied to wind-specific AI models, so that it becomes easier to explain why something was predicted or why a certain choice was given.

– People should be allowed to interact directly with AI and give feedback when necessary, so they are sure about the results and can correct any errors.

4.10.3. *Digital twins and physics-informed AI*

AI can perform poorly for tasks that it was not trained for or for rare cases. Having physical laws in the model increases its strength and wide-ranging use.

Research directions:

– High fidelity digital twins depend on the regular input from sensors and physical models to learn from the operations and support special case studies for how the system should be maintained and managed.

– PINNs make use of governing equations such as Navier–Stokes to help with prediction and to minimize the amount of data needed.

4.10.4. *Multi-agent and cooperative control*

Centralized control becomes hard to maintain and may cause a critical failure as wind farms increase in size.

Research directions:

– MARL algorithms should be designed so that each turbine can work independently while cooperating with others to achieve things such as managing the wake and equal load sharing.

– Centralized control of the farm should be combined with fast controls on the turbines to ensure that the main goals of the project are met but the local characteristics of the installation are still considered.

4.10.5. *Integration with energy storage and multi-energy systems*

Wind variability can be tackled by using storage (batteries and hydrogen) together with demand side resources.

Research directions:

– AI models can be created to help plan how much wind energy to generate, store and adjust in demand to create high profits and ensure grid stability.

– Wind farm digital twins can be used along with data for energy storage and interactions with the grid, to look at energy management from all sides.

4.10.6. *Sustainable AI and green computing*

Training and using AI in large quantities require a lot of energy and add carbon emissions, which can cancel out the positives regarding wind energy.

Research directions:

– Developing AI models and computing approaches that are very efficient and only need to work locally.

– It is helpful to conduct AI training when renewable energy is being produced in greater amounts, as this reduces the carbon impact.

4.10.7. *AI for materials and recycling optimization*

The fact that wind energy is sustainable includes its creation, directed sourcing of materials and end-of-life recycling.

Research directions:

– Generating models for materials: variational autoencoders and similar generative models can suggest useful blades that are easier to recycle and avoid fatigue.

– Thanks to machine learning, models can optimize the timing of taking devices out of use and the routes for recycling to help lower environmental damage.

4.11. Conclusion

With the help of AI, it becomes possible to sustain wind energy throughout the entire process, including in initial site exploration, resource forecasting, everyday operation, upkeep and concluding its operation.

With AI's ability to analyze a lot of different types of data and learn difficult patterns, wind farms can be improved.

1) Resource productivity is increased: because of machine learning models such as RF and Gradient Boosting, wind speed and power forecasts over terrain have improved a lot, yielding better estimates of energy and improving how turbines are arranged.

2) Losses due to wakes are minimized and control is improved: applying wake steering and MPC based on reinforcement learning, optimal adjustments in yaw and pitch are made to lessen wake disturbances and boost production at the farm level by approximately 3–5% with a reduction in the loads on the turbine.

3) Better prediction and increased reliability mean an increase in uptime: the use of LSTM networks and GPR for finding irregularities and time remaining only increase turbine availability and allow lower maintenance expenditures.

4.12. References

Abadi, M., Chu, A., Goodfellow, I., McMahan, H.B., Mironov, I., Talwar, K., Zhang, L. (2016). Deep learning with differential privacy. In *Proceedings of the 2016 ACM SIGSAC Conference on Computer and Communications Security*, Vienna, 308–318.

Barthelmie, R.J. and Jensen, L.E. (2010). Evaluation of wind farm efficiency and wind turbine wakes at the Nysted offshore wind farm. *Wind Energy*, 13(6), 573–586.

Cadenas, E. and Rivera, W. (2010). Wind speed forecasting in three different regions of Mexico, using a hybrid ARIMA–ANN model. *Renewable Energy*, 35(12), 2732–2738.

Chen, T. and Guestrin, C. (2016). Xgboost: A scalable tree boosting system. In *Proceedings of the 22nd Acm Sigkdd International Conference on Knowledge Discovery and Data Mining*, San Francisco, CA, 785–794.

Ding, T., Sun, Y., Huang, C., Mu, C., Fan, Y., Lin, J., Qin, Y. (2022). Pathways of clean energy heating electrification programs for reducing carbon emissions in Northwest China. *Renewable and Sustainable Energy Reviews*, 166, 112679.

Ding, S., Yang, C., Zhang, S. (2023). Acoustic-signal-based damage detection of wind turbine blades – A review. *Sensors*, 23(11), 4987.

Field, R.D., Luo, M., Bauer, S.E., Hickman, J.E., Elsaesser, G., Mezuman, K., van Lier-Walqui, M., Wu, J., Tsigaridis, K. (2023). Estimating the impact of a 2017 smoke plume on surface climate over northern Canada with a climate model, satellite retrievals, and weather forecasts. *J. Geophys. Res. Atmos.*, 129(15). doi: 10.1029/2023JD039396.

Fox, H., Pillai, A.C., Friedrich, D., Collu, M., Dawood, T., Johanning, L. (2022). A review of predictive and prescriptive offshore wind farm operation and maintenance. *Energies*, 15(2), 504.

Hadi, M.B., Moeini-Aghtaie, M., Khoshjahan, M., Dehghanian, P. (2022). A comprehensive review on power system flexibility: Concept, services, and products. *IEEE Access*, 10, 99257–99267.

Hong, T. and Fan, S. (2016). Probabilistic electric load forecasting: A tutorial review. *International Journal of Forecasting*, 32(3), 914–938.

Jeon, T., Kim, D., Paek, I. (2021). Design and analysis of yaw command tracking control algorithm for wind turbine for wind farm control application. *Journal of the Korean Solar Energy Society*, 41(3), 65–78.

Lund, H. and Mathiesen, B.V. (2009). Energy system analysis of 100% renewable energy systems – The case of denmark in years 2030 and 2050. *Energy*, 34(5), 524–531.

Márquez, F.P.G., Tobias, A.M., Pérez, J.M.P., Papaelias, M. (2012). Condition monitoring of wind turbines: Techniques and methods. *Renewable Energy*, 46, 169–178.

Mehrabi, N., Morstatter, F., Saxena, N., Lerman, K., Galstyan, A. (2021). A survey on bias and fairness in machine learning. *ACM Computing Surveys (CSUR)*, 54(6), 1–35.

Meinshausen, N. and Ridgeway, G. (2006). Quantile regression forests. *Journal of Machine Learning Research*, 7(6), 983–999.

Mohammadi-ivatloo, B. and Nazari-Heris, M. (eds) (2019). *Robust Optimal Planning and Operation of Electrical Energy Systems*. Springer, Cham.

Peng, H., Zhang, H., Fan, Y., Shangguan, L., Yang, Y. (2022). A review of research on wind turbine bearings' failure analysis and fault diagnosis. *Lubricants*, 11(1), 14.

Raissi, M., Perdikaris, P., Karniadakis, G.E. (2019). Physics-informed neural networks: A deep learning framework for solving forward and inverse problems involving nonlinear partial differential equations. *Journal of Computational Physics*, 378, 686–707.

Shi, X., Chen, Z., Wang, H., Yeung, D.Y., Wong, W.K., Woo, W.C. (2015). Convolutional LSTM network: A machine learning approach for precipitation nowcasting. *Advances in Neural Information Processing Systems*, 28, 802–810.

Tang, X., Heng, J., Kaewunruen, S., Dai, K., Baniotopoulos, C. (2024). Artificial intelligence-powered digital twins for sustainable and resilient engineering structures. *Bauingenieur*, 99(9), 270–276.

Veale, M. and Edwards, L. (2018). Clarity, surprises, and further questions in the Article 29 Working Party draft guidance on automated decision-making and profiling. *Computer Law & Security Review*, 34(2), 398–404.

Vidal, O., Goffé, B., Arndt, N. (2013). Metals for a low-carbon society. *Nature Geoscience*, 6(11), 894–896.

Williams, C.K. and Rasmussen, C.E. (2006). *Gaussian Processes for Machine Learning*. MIT Press, Cambridge, MA.

Wolsink, M. (2007). Wind power implementation: The nature of public attitudes: Equity and fairness instead of 'backyard motives'. *Renewable and Sustainable Energy Reviews*, 11(6), 1188–1207.

Xiong, J., Ye, Y., Wang, Q., Dong, X., Lu, T., Ma, D. (2024). A comprehensive review on distributed energy cooperative control and optimization method for energy interconnection system. *Electric Power Systems Research*, 237, 111007.

Xu, D., Shao, H., Deng, X., Wang, X. (2022). The hidden-layers topology analysis of deep learning models in survey for forecasting and generation of the wind power and photovoltaic energy. *Computer Modeling in Engineering & Sciences*, 131(2), 567.

5

Intelligent Energy with AI-Driven Innovations in Wind Power Systems

Using AI in wind power is completely changing the way renewable energy is generated and organized. The purpose of this study is to improve the efficiency, dependability and output quality of wind turbine operation with AI tools which are then simulated in MATLAB. Designing, running simulations on (and validating algorithms related to) wind turbine control, fault detection and power output is made possible with MATLAB. With machine learning, it is now possible to forecast wind energy and speed very accurately which helps improve the grid's stability. Using Simulink models, the behavior of the turbine is shown under a range of changing wind speeds. Thanks to AI, machine malfunctions can be found early, helping to save time from unexpected outages. Fuzzy logic and neural networks are introduced to ensure that pitch angle and generator torque are set right for excellent power extraction.

The test results highlight important increases in energy efficiency and reliability of operation. These decision-making models are approved through checking with real data to confirm they are reliable. AI is used in the method to forecast loads and help manage energy demand. By reducing energy loss and using resources more wisely, we observe good environmental results. The simulation results suggest that AI can help wind power systems get smarter and more adaptable. The results demonstrate that AI plays a key role in the development of new sustainable energy solutions.

5.1. Introduction

With the rising importance of sustainability and carbon neutrality around the world, there is a faster move toward renewable energy and wind power is gaining attention as a key option. Still, the tendency of wind energy to change and stop

Chapter written by R. RAJASREE, D. LAKSHMI and Malathy BATUMALAY.

working makes managing and improving large-scale use of wind challenging (Dhayal et al. 2025). As a result, wind power systems are combining artificial intelligence (AI) for better control, improved forecasts and automated maintenance. Using AI such as machine learning, neural networks and fuzzy logic helps systems to identify complex patterns and take fast, smart decisions which greatly boosts the stability and outcomes of wind energy systems.

AI is currently used to make energy systems more intelligent, allowing them to automatically handle the effects of the environment, run turbines better and anticipate mechanical failings. The use of sensor activity, weather details and earlier operation history help AI improve the assessment of wind energy, expectations for its yield and the stability of the grid. Repairs and maintenance are reduced, as well as raising the energy efficiency of the entire system. Integrating AI into wind energy systems is an important move forward for making our energy structure more sustainable, efficient and smart (Elmousalami et al. 2025).

5.1.1. *AI in renewable energy*

AI is changing the renewable energy sector by supporting better, faster and more efficient management of generation, distribution and consumption. It is often hard for old energy systems to deal with the unpredictable nature of wind and solar energies. To deal with these problems, AI manages a lot of data, finds patterns and updates operations rapidly to improve the system's performance. Experts in the energy sector use machine learning, neural networks and fuzzy logic to predict energy usage, run control processes and help stabilize the grid as energy levels vary (Elmousalami et al. 2024).

For renewable energy applications, AI is important for improving how we predict and use energy, manage power production and save money on operations. It allows predictive upkeep of infrastructure, which means less time offline and longer life for equipment. AI technology is applied in smart grids, where it looks after scattered energy resources, predicts demand and guarantees power stability and strength (Alnaser et al. 2024). AI systems with renewable energy are more flexible, smart and able to handle modern energy needs, helping to make the change to a sustainable global energy system.

5.1.2. *Wind resource assessment and forecasting using AI*

Checking the wind resource is necessary before starting the development of a wind energy project. To judge if a wind site is suitable for energy generation, professionals assess its wind speed, direction and frequency. Assessment in this area meets heavy reliance on long-term data and models, making the task slow, expensive and with a risk of error because it is not easy to model the detailed changes in atmospheric behavior. By making use of enormous data and effective learning techniques, AI provides a quicker and more accurate way to predict how wind influences power systems (Elmousalami and Sakr 2024a, 2024b). Thanks to AI and especially to machine learning and deep learning, forecasting wind with AI relies on historical winds, current weather and topography to calculate fine-grained predictions. Short- and long-term wind predictions frequently use support vector machines, Random Forests and neural networks. Thanks to these models, the program can learn from ongoing experience and get better the more it is trained.

Moreover, AI uses imagery from satellites, readings from sensors and climate data to precisely analyze wind in every part of a site, allowing developers to make more accurate wind energy planning decisions. Precise wind forecasts made possible by AI help with energy production, keep the grid stable and improve the economy. By relying on forecasts from technology, grid operators can adjust energy supply and use, which means they need less fossil fuel during emergencies (Cakiroglu et al. 2024).

Meanwhile, improved forecasting lowers the amount of wind power curtailed, makes the best use of storage units and increases the grid's stability. With the improvements AI brings in assessing wind, there are fewer risks for wind energy investors and operators, helping drive the growth of cleaner energy systems.

5.1.3. *AI-based control and optimization of wind turbines*

With the help of AI, power companies can now manage and optimize wind turbines easily in response to weather-based conditions. Using traditional methods, when facilities react to situations the same way every time, the equipment can be damaged, resulting in inefficiency. Unlike traditional systems, AI algorithms such as neural networks, fuzzy logic and reinforcement learning use sensor measurements from turbines to automatically alter control settings. Making these adjustments helps set the blade's angle, align the yaw mechanism and control generator power, which leads to better energy capture and less damage to important parts (Cruz et al. 2024).

AI benefits wind farms by managing turbine activity and reducing interference, known as wake effects, from upstream turbines on those downstream. Machine learning models can address these aerodynamic matters, guaranteeing the same power output throughout the farm. Moreover, these systems can manage multiple goals at once, for example, boosting output, easing the impact on machines and setting up maintenance times based on information from previous and current data (Elmousalami and Sakr 2024a, 2024b). So, AI makes wind energy more efficient, reliable and economical to operate alone.

5.1.4. *Predictive maintenance and fault detection in wind systems*

Wind turbines are built to operate in difficult environmental conditions, which increases their chances of damage or failure. The usual forms of maintenance such as those done on an appointed or emergency timetable are often both costly and inefficient. Such methods may bring unexpected delays or responses to urgent problems. When AI is used for predictive maintenance, organizations can work ahead and address problems with more ease. By always checking the health of wind turbines and analyzing live data, AI systems can forecast issues, which allows action to be taken early and where it is most needed. Predictive maintenance for turbines brings in machine learning, neural networks and information based on sensor data in the gearbox, blades, bearings and generator (Fukutomi 2024). They use information on normal behavior to spot the early signs of something going wrong.

In addition to finding the possibility of a fault, AI also improves fault classification and root-cause discovery, so operators find out when, where and why a failure could happen. With AI and IoT, operators can monitor all parts of a wind farm, whether it is close to people or far out in the ocean. Reliance on digital tools boosts reliability, brings down costs and provides better direction for asset decisions (Karijadi et al. 2023). As wind power increases in size and difficulty, using AI to predict and maintain problems plays a key role in achieving high operation, safety and profitability of solar and wind plants.

5.1.5. *Smart grid integration and load management*

Wind integration into the grid is unique because wind flows can change and suddenly stop. Like other clean energies, wind does not work like a regular energy source and has to be managed carefully so the grid remains stable. AI in smart grids provides a more efficient way to bring wind energy into the power system. AI-powered systems review data from the grid right away, project energy use in advance and organize power from several sources. AI is important for computing energy load predictions and controlling consumption on the demand side in smart grids (Malakouti 2023). With help from machine learning, firms in the energy industry can estimate future energy use by looking at past data, the weather outlook and what consumers usually do. Thanks to this data, electricity stored by the grid can be used efficiently, reducing fossil fuel backups and ensuring that wind energy is mostly used when it is most available.

AI is also used to link power grids of different parts into unified networks that allow home-based energy resources such as solar panels, wind turbines and batteries to manage their energy needs effectively. Grids with these components will coordinate local energy use and generate it whenever and wherever needed. With AI, the handling of both energy providing and using becomes better, which helps the infrastructure become stronger, more adaptable and greener (Tornede et al. 2023). Using AI to operate wind power and smart grids is important for preparing a new energy system to meet the challenges ahead.

5.2. Literature review

There has been high carbon emission from the industry for many years, so we are now concentrating on I5.0 for sustainable progress. Through the green solutions discussed here, production can support the principles of I5.0 and become more sustainable. In this study, the literature from 2012 to 2024 is analyzed with Scopus to examine what AI, green supply chains and the metaverse do. It points out the crucial reasons why industry sustainability is necessary. Advantages and disadvantages help to influence decisions on policy and technology. Using these results, policymakers, researchers and people interested in the environment can make better decisions (Dhayal et al. 2025). For this reason, everyone has a role to play in making a carbon emission-free world.

To be managed properly, renewable energy must be supported by correct forecasts of wind speed and power. This research introduces SAI-WEFS, an AI-powered system for forecasting emissions that was built from a real scenario in the MENA region. The study examines twelve machine learning techniques, single ones and part of ensembles, for data from multiple time horizons. Wind speed predictions are used within the system's power prediction models. Environmental impact measures the amount of CO_2 produced by computing during an hour. Both of these approaches delivered higher accuracy and lower emissions. The Urban Carbon Emission Index is predicted by SAI-WEFS to assist with urban energy decarbonization (Elmousalami et al. 2025).

Renewable energy systems depend greatly on accurate predictions of wind speed and power (Elmousalami et al. 2024). In this analysis, twelve ML and DL algorithms are assessed, particularly groups of ensemble models for timeframes of up to 36 hours. A suggestion is made to use wind speed forecasts to drive the wind power modeling process. Accuracy of extremely random trees and bagging approaches was high, with MAPE measures from 3.4% to 11.7%. MSE and adjusted R^2 were used as the evaluation metrics. It was easy to train extreme gradient boosting because it did not take much computing time. Air pressure, wind vane and humidity were found to be the most important predictors by sensitivity analysis.

This study looks at the ways AI-based digital twins and IoT help achieve sustainable building designs. The research includes analysis of 125 scientific articles based on four major themes. First, it looks at digital twins used in construction, building management and smart cities. The study also covers advances in automation and AI in the Industries 4.0 and cyber-physical systems. Third, it discusses new areas such as blockchain, cybersecurity and EEG-driven systems in how cities are built (Alnaser et al. 2024). Methods that depend on data for flood resilience and modern urban systems are also explored. The assessment demonstrates that digital solutions and AI help make urban and industrial systems more sustainable and efficient with energy.

Under difficult drilling conditions, lost circulation results in 10–20% of total costs. An Automated Lost Circulation Severity Classification and Mitigation System (ALCSCMS) is proposed in this study. Data from 65,377 drilling operations was used to design and enhance eleven ensemble models using Bayesian techniques (Elmousalami and Sakr 2024b). There were no misclassifications using the Random Forest model. With a genetic algorithm, machines are led to the correct decision by turning severe cases into acceptable ones. Severity is ranked into five levels, and the model uses optimized input values to decrease risks. SHAP analysis shows which drilling factors play a big role in creating lost circulation situations.

Wind energy is important, but the high cost means it is vital to estimate its usage before starting work. The research used six machine learning models to predict how much power wind turbines generated in Canakkale, Turkey, between 2011 and 2020, based on environmental data. The authors used XGBoost, Random Forest, LightGBM, CatBoost, AdaBoost and M5-Prime algorithms. To explain the predictions of the model, SHAP analysis was used. The highest R^2 accuracy in this case was achieved by XGBoost, and LightGBM was shown to require the least time for computation. It was determined that wind speed had the greatest influence during classification, among all features. The research suggests that using data can guide effective wind power planning (Cakiroglu et al. 2024).

Forecasting wind speed and power correctly is essential for making renewable energy more efficient in smart and zero-energy cities (Cruz et al. 2024). This study presents SAI-WEFS, an environmental AI forecasting system developed using a case in the MENA region. Over the course of multiple timeframes, it looks at twelve machine learning algorithms, some as single methods and others as grouped models. Wind power forecasts depend on forecasts of wind speed. Environmental impact is assessed by checking the CO_2 emissions from a computing job per hour. The best results were achieved by ensemble trees, which produced lower emissions than deep learning models. Urban decarbonization is supported by SAI-WEFS through the estimation of the Urban Carbon Emission Index.

In extreme circumstances, 10–20% of drilling costs are caused by lost circulation and mud losses. ALCSCMS is the system introduced in this chapter to automate lost circulation severity (LCS) prediction and control. With Bayesian methods, we created and improved eleven ensemble machine learning models using 65,377 data points (Elmousalami and Sakr 2024). Random Forest was fully accurate on the test set. Genetic algorithm was applied to turn the most extreme LCS classes into acceptable levels by making changes to input parameters. Cyber threats are organized into five groups, and the system works to improve the way the threats are managed. SHAP analysis allowed us to identify main input factors controlling how LCS is predicted.

Because of where they are and the wealth of their resources, the Middle East and North Africa continue to see much conflict. We therefore explore three main reasons behind persistent instability: colonial rules, Cold War conditions and economic features. How Britain, France and Italy were once in charge of these regions changed society permanently. Because the region lay within the borders controlled by the U.S. and Soviet blocs, it only became more unstable during the Cold War. Important aspects are a weak manufacturing sector, a high unemployment rate and not having a democratic government. Energy independence in the U.S. since the 2000s has involved changing its oil focus and boosting its shipment of arms to the Middle East. Middle Eastern crises are still being driven by these economic and political factors (Fukutomi 2024).

Reliable grid operation in renewable energy systems greatly depends on accurate wind power forecasting. Because the wind changes without warning, making a precise forecast is challenging. In this study, CEEMDAN, EWT and LSTM are combined to develop a method for ultra-short-term wind power forecasting. CEEMDAN splits the data into several subseries, and EWT reduces the effects of high-frequency noise in the data (Karijadi et al. 2023). Each subseries is forecasted using an LSTM model, and the final predictions are made by combining all of the results. The outcomes of this method are validated using data collected in France and Turkey. The test results confirm that they are more accurate than known forecasting techniques.

As fossil fuel power plants cause global warming, this study examines wind energy as a possible alternative. Our research focused on a wind farm in Texas. The models Random Forest, AdaBoost and K Neighbors were used to predict both the turbine power and the wind's speed. ME, MAE, MAPE, RMSE, R^2 and runtime were all included in the evaluation metrics used. Both wind speed and power predictions were most accurate when using Random Forest. The memory algorithm K Neighbors scored poorest, yet executes faster than the others. Research enables us to understand that wind energy helps reduce damage to the environment (Malakouti 2023).

By using AutoML, users can automate the creation of ML pipelines, while some are worried about its need for a lot of energy and resources. This chapter therefore proposes using Green AutoML to make AutoML processes more environmentally friendly (Tornede et al. 2023). It initially shows how energy and carbon costs of AutoML systems can be measured. After that, it explores ideas for making AutoML systems more sustainable through their design and evaluation. The study calls for honest reporting on how the environment is affected. It also suggests giving scientists rewards to encourage greener solutions. A checklist for sustainability is used to guide new AutoML research in an eco-friendly direction.

As AI use increases, so does its effect on the planet; and to handle this problem, Green AI is growing. After examining 98 Green AI studies, this chapter highlights strong growth in the field since 2020. Much of the research is aimed at keeping track of how models make predictions, tuning the applications' parameters and checking the results against previous examples. Research works usually ignore the type of algorithm, make heavy use of pictures and highlight the process of training. Most research is done in the laboratory, and experiments have shown energy savings of up to 115% (Verdecchia et al. 2023). Whereas the industry is showing interest, the main focus of work is on academia and tools are hard to get. The study finds that Green AI is maturing and can now be used more widely in industries.

The way cities in the U.S. use building energy in the future will depend on climate, new technologies and social and economic factors. Using a detailed bottom-up method, this study mirrors a mid-century hourly energy use in 277 urban areas. Tests indicate that warming can cause average peak electricity use intensity (EUI) to increase by 10.1% to over 37% and in extreme cases over 110%. Heat demand goes down by approximately 10% with every 1°C temperature rise and cooling demand rises by about 14%. Many factors such as a city's population and its changing power sector shape the range of energy use in cities (Wang et al. 2023). We find that as temperature changes (-2.5% to -2.0%), an increasing world population (up to 52.2%) and decarbonization (up to 17.1%) affect electricity use. They show that local energy plans are essential.

To make efficient use of renewable energy in smart and zero-energy cities, we need accurate wind speed forecasts and estimates of available wind power. This work presents SAI-WEFS, an AI-led low-carbon weather forecasting system based on data from the real MENA region (Yaghoubirad et al. 2023). Twelve machine learning models, including both ensemble and single models, are examined in this experiment at several key time points. Inputs from wind speed forecasting make power prediction models more accurate. Emissions of CO_2 are the main factor used to measure environmental impact in computing. Other things being equal, tree-based ensemble models perform better and are more reliable than most alternatives. To support the Urban Energy Supply Decarbonization Framework, SAI-WEFS computes the Urban Carbon Emission Index.

Wind energy helps support both sustainable growth and the aim of making energy cleaner (Wu et al. 2022). Wind resource and power forecasting is necessary for stronger adoption of wind power, but it is still a difficult task because of its unpredictability. Wind forecasting is now commonly supported by the use of innovative deep learning solutions. In this chapter, we review how RNNs, restricted Boltzmann machines, CNNs and auto-encoders are applied in wind energy. Such models provide better results when analyzing data that keeps changing over time. Future directions for forecasting with deep learning are also considered in the report. It points out that AI is playing a bigger part in helping wind energy systems advance.

Lately, the fast development of AI has made people consider AI sustainability. For this study, we used systematic mapping to examine publications connected to economic, social and environmental topics. Research results point to a change from isolated studies to a full grasp of AI sustainability. The discipline now covers both helping sustainability with AI and improving AI's own environmental impact. Researchers are exploring issues by looking at them from multiple perspectives. The chapter supports an open-minded and detailed look at the topic. For this subject to advance, it must be supported by various fields early on (Wilson and Van Der Velden 2022).

Wind power has grown rapidly because it supports efforts to cut greenhouse gas emissions and fight climate change. Correct forecasting is necessary for successful operations on the grid and on energy markets. Because the wind is unpredictable, traditional approaches find it hard to manage wind energy and keep operations consistent (Malakouti et al. 2022). This research introduces machine learning methods to make wind power prediction more accurate. The algorithms applied are extremely randomized trees, Light GBM, various ensemble methods and CNN-LSTM. Our evaluation demonstrates that CNN-LSTM obtained the lowest mean square error and the most accurate results. These techniques are also efficient because they are easy to compute. They help wind power companies make better decisions.

Increased electricity use, lowering fossil fuel use and increased carbon emissions are just a few of the growing problems in power systems. RES such as wind in our energy mix is necessary for building a sustainable future. Over 850,000 measurements from a wind farm in 2020 are analyzed in this study. The performance of the wind farm was predicted using machine learning techniques. A selection of algorithms was tested, including Extra Trees, LightGBM, Gradient Boosting Regressor, Decision Tree, AdaBoost and Ridge. Extra Trees were able to make the most accurate predictions. Ridge performed the worst of the lot. The findings show that RES can help make power systems greener (Malakouti and Ghiasi 2022).

The increase in wind power integration creates more problems for planning the power grid due to unpredictability. This research introduces a framework for temporally predicting wind power using ERA5 reanalysis data, LightGBM, MIC and nonparametric regression (Liao et al. 2022). ERA5 offers a wide range of meteorological information, and MIC picks out the important features. The tool can accurately predict trends in operations data from the past. Interval prediction is made possible by nonparametric regression, which increases the usefulness of forecasts. Effectiveness of the model is determined by using CORR, RMSE, MAE, IA and KGE. The framework is more accurate than SVR, RF and XGBoost. Energy interconnection increases safety and dependability of the grid.

In this chapter, we study how fast-growing AI influences our surroundings by analyzing all parts of AI, including its data, algorithms and hardware. The study checks the emissions caused by AI from the initial design phase to applying it on a large scale. The study looks at emissions that happen because of both working and making AI devices. A full analysis of the process highlights that combining hardware and software designs can lower greenhouse gas emissions (Wu et al. 2022). Industry challenges and insights are provided to guide businesses toward better use of AI. The chapter calls for environmentally friendly practices in the field of AI development. It describes further research that needs to be done to achieve this aim. The idea behind the findings is to encourage growth in sustainable AI.

Sustainable AI, the main focus of this chapter, requires AI projects to consider nature and justice at every stage of their operations (Van Wynsberghe 2021). When we talk about sustainable AI, we are looking at the system and the wider consequences for society. It encourages developments in AI that match with caring for the planet, saving the environment and giving everyone fair access to benefits. The subject covers two fields: AI for supporting sustainability and sustainability in AI, for instance by not increasing carbon emissions. The chapter points out the key disputes between progress in AI and making sure resources are shared justly across generations and among environment, society and the economy. Its purpose is to encourage people involved in AI to see the impact it has on the environment. It depends, therefore, on the creation of policies and allocation of financial resources for sustainable AI.

This chapter looks closely at net-zero carbon cities by outlining their purposes, features, progress and performance. It sets net-zero apart from low-carbon cities and explores their relationship to account for greenhouse gases and deep decarbonization. The research explores urban infrastructure, planning methods and the challenges met in putting them into action. To succeed with net-zero, we must affect many sectors and also take up carbon sequestration measures. Our examination of ongoing implementation efforts has pointed out achievements as well as areas that still need work (Seto et al. 2021). It is necessary to put mitigation efforts in the proper order, because of both carbon lock-in and the way people behave. For the success of urban projects, plans for growth and development must fit with the layout of cities. What is becoming clear from the review is that net-zero cities can exist, but only after transformative changes are made to planning.

There is little agreement on how to measure carbon at the city level and structures for studying connections between cutting emissions and following the sustainable development goals are minimal (Ramaswami et al. 2021). A new approach for accounting that uses key physical support systems can help fill these knowledge gaps and contribute to urban transitions.

It is important for urban systems to coordinate across sectors to make serious progress on climate mitigation. A policy framework with the SDEWES index is used in this study to review the performance of urban areas. The description and analysis of unintegrated, emerging and advanced stages of integration are presented. A case study municipality received a score of 27.783 and ranked 79th out of 120 cities. Action plans that exist are reviewed against actions proposed to enhance integration and index results. Experiences from other cities encourage cities worldwide to act together on climate change. The results emphasize that achieving net-zero targets depends on bringing together various aspects of the system. These models help find ways to build urban energy systems that rely solely on climate-neutral and renewable energy (Kılkış 2021).

A proper forecast helps make decisions about how to run the electrical grid more efficiently. An artificial neural network approach is used in this study to short-term predict wind speed, solar radiation and power demand. A NARX network is assessed as a competitor to existing modeling techniques. It is preferred when making forecasts for far-ahead periods of 8–24 hours (Di Piazza et al. 2020).

Sustainability simulations indicate that the model works well for short-term predictions. An important benefit is that it is relatively straightforward and can be done at low expense. The model is designed so that performance and efficiency work at the same level. It provides a straightforward answer for energy forecasting use cases.

This study examines how cost prediction methods use both qualitative and quantitative techniques as well as those based on computational intelligence (CI). Twenty different AIs, including ANN, SVM, RF, XGBoost and GA, are checked to create cost models (Elmousalami 2020). An important part of the book shows how different models and trends can be used in practical recommendations. From the results, the study concludes that XGBoost is the most precise way, with a MAPE of 9.091% and an adjusted R^2 of 0.929. Nonlinearity, missing data and uncertainty in models are problems covered. An actual field example for field canal improvement projects (FCIPs) is discussed. This chapter provides a guide for cost modeling at the beginning of any project. ChemRXiv gives free access to a public dataset for anyone interested in conducting further research.

5.3. Proposed methodology

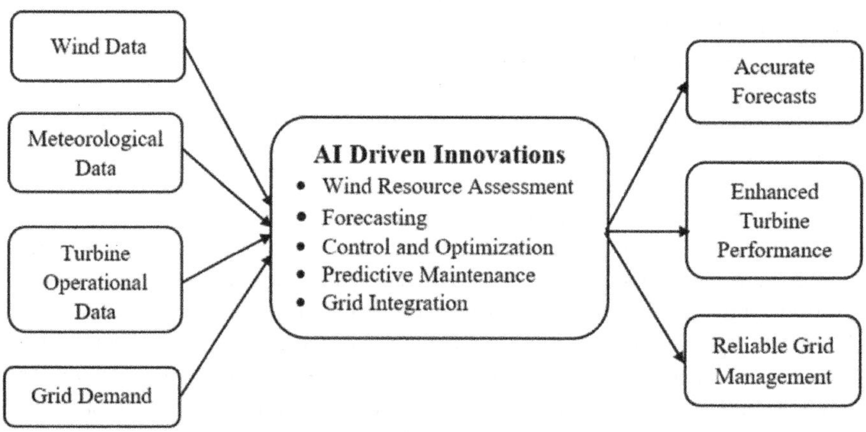

Figure 5.1. *Flow diagram of AI-based predictive maintenance system*

Figure 5.1 shows a breakdown of the ways in which AI improves the performance and stability of wind farms. The left side of Figure 5.1 lists four main data sources: wind, meteorological, operational turbine and grid demand details. The data on wind shows both recent wind conditions and past records of wind speed and direction, both of which are important for judging if wind energy will be productive and for making forecasts. Information about temperature, humidity and pressure is important in understanding how wind patterns and how efficiently wind turbines operate (Verdecchia et al. 2023). This gives us the necessary data such as blade pitch, rotor speed, torque and energy output for supervising the turbine on a routine basis while maintaining it well in advance. Electricity consumption trends indicated in the grid demand data make it easier for supply and demand to be well matched. Once we supply these inputs into an AI-driven system, powerful algorithms look at the data and handle tasks important for energy management.

AI-driven innovations are featured at the heart of the diagram, made up of wind resource assessment, forecasting, control and optimization, predictive maintenance and grid integration (Wang et al. 2023). AI models are used to study the likely future performance and wind generation at specific sites. With machine learning, wind energy operators can forecast short- and long-term production and prepare accordingly. Adjustments according to information collected from the environment and use allow the system to work at its best. With predictive maintenance, businesses are able to stop equipment problems early, keep machines running more and save on maintenance. Grid integration ensures system operators can evenly share the developed wind power as needs on the grid change. Results from these steps are given on the right: more accurate predictions, improved performance of the turbines and confident handling of the grid (Yaghoubirad et al. 2023). The studies demonstrate the importance of AI in helping wind energy systems become more efficient, sustainable and in line with the goals of modern energy and a decarbonized society.

5.3.1. *Wind data*

A wind power system depends heavily on wind data, made up of live and historical wind speed, direction, turbulence and gust measurements (Wu et al. 2022). Most of the time, this data is gathered using anemometers, wind vanes, LiDAR systems and remote sensing technologies. Having precise wind information is necessary to determine if a site is suitable and to estimate the power that wind turbines could produce.

Azimuth is very important for setting up AI forecasting to predict changes in wind for both quick and long periods. Since wind can change quickly, energy companies must use fine-detailed wind information from space and time to deal with the uncertainty. Having wind data allows the turbine to be adjusted as needed to increase energy production. It also helps energy planners determine the capacity values of turbines and find the perfect placement for each. If wind data is not accurate, all further processes, for example, control, forecasting and upkeep, are affected. It helps the grid by projecting the wind-related part of the electricity supply. It is necessary to study wind cycles during the day so you can expect changes in energy production. AI systems have to process a lot of data to increase how well they learn. The information is regularly sent into simulations and optimization software (Wilson and Van Der Velden 2022). Overall, wind data is what supports intelligent wind energy systems.

5.3.2. *Meteorological data*

Meteorological data includes important environmental elements such as temperature, humidity, pressure, solar radiation and precipitation. Wind formation, air density and turbine effectiveness are all affected by these elements (Malakouti et al. 2022). Wind speed depends on warm air and high pressure, which influences how windmills create electricity. Models using meteorological and wind information together provide more reliable weather predictions. It allows us to see ahead of any weather events that might increase risks to or reduce the output of turbines. In certain situations, when storms or ice are detected, meteorological sensors kick in and perform a shutdown. Meteorological information plays a key role in improving and refining automated weather predictions. The information is also used to estimate energy output over a longer term and enhance the assessment of feasible projects. Climate data helps in understanding wind data by showing the wider environmental surroundings. It allows us to study how external factors modify the way a turbine works inside. Machine learning makes it possible for AI to uncover how environmental and mechanical systems are connected (Malakouti and Ghiasi 2022). Data from weather stations at the moment is used for maintaining assets based on conditions and handling grid electricity loads. That is why AI systems are better at adapting to the changing situations in wind energy production.

5.3.3. *Turbine operational data*

Within the dataset from turbines, we can find aspects such as the blade pitch angle, the rotor's speed, the energy output, torque, temperature and vibration levels. Data from within the turbines is needed to ensure they stay healthy, efficient and safe (Liao et al. 2022). They make use of data to learn the usual way things operate and find any irregularities, which helps with preventive maintenance. An example is that if vibration goes up quickly, it may point to either a problem with blade balance or damage of the bearings. Maintenance teams can be alerted by these models to avoid costly issues from occurring. Using the information from operational data, turbine parameters can be optimized to meet the present conditions. In combination with other data, it allows controllers to adjust the operating speed of turbines in real time to boost energy generation. Approaching your tools with constant, real-time checks pays off by increasing their operational life and reducing the risk of a breakdown. Operational logs are looked over by machine learning algorithms to discover flaws and recommend controls. We use the data to design digital twins of turbines, testing different strategies to make them work better. How turbines and wind farms perform operationally matters a lot when comparing results (Wu et al. 2022). The details from data help inform actions by engineers and AI. When all aspects of turbine operation are analyzed, wind energy becomes safer, more intelligent and dependable.

5.3.4. *Grid demand*

To measure grid demand, we look at real-time meters and SCADA systems, which give the total amount of electricity being consumed each minute. It changes depending on the weather, the time of day, what happens in the economy and how consumers act (Van Wynsberghe 2021). To prevent waste or shortages, wind power systems must create electricity at times when the grid needs it. Helped by AI, the wind turbine output can change or energy storage times be set based on demand predictions. The result is a stable grid and prevents cases of overloading or under-voltages. Data on customer demand is needed to manage the smart grid, including methods that move load according to the electric supply. By having a clear forecast of demand, companies can join the market by deciding how to bid and making their strategies according to likely energy prices. With grid demand data, AI estimates the energy that should be either added to (or taken from) the grid in real time. The information enables utilities to stabilize the energy supply from wind with the minimum load needed from other energy sources (Seto et al. 2021). During high usage times, understanding demand lets us make better use of the turbines and send out the right amount of energy. Such models allow us to predict demand levels for different climate changes or rules. Basically, it is unnecessary to say that demand data is necessary to align wind energy with the demands and movements of the grid. It makes it smooth for variable renewable energy to work together with existing energy grids.

5.3.5. *AI-driven innovations*

All incoming data is first analyzed by AI computation methods at the heart of the AI-driven innovation block. It applies machine learning, deep learning, ensemble methods and occasionally reinforcement learning to inspect, understand and handle issues in wind energy systems. Wind resource assessment is used here to determine the potential energy, and forecasting is used to predict wind speed and turbine output for particular durations (Ramaswami et al. 2021). Real-time control and optimization are possible through altering blade pitch, yaw and generator settings according to the environmental situation. By catching small problems early with data, predictive maintenance makes machines last longer and keeps work areas safe. It also plays a key role in grid integration, making certain generated electricity fits demand and makes grid electricity delivery more stable. Because the system uses a wide variety of information, it keeps improving and stays efficient as it is updated with fresh data. AI is used to create simulations to assist in future plans and to adjust turbine settings. This area of the computer is also the digital twin's brain, copying what a turbine does in a simulation instead of the real world. It leads to greater energy generation, better running of machines and lowers expenses. For the most part, AI-driven innovations are responsible for turning wind systems from simple to autonomous and smart.

5.3.6. *Accurate forecasts*

This means the system is able to predict wind and resulting power values with great precision. It involves giving predictions that are close in time (up to hours) and predictions that are far in time (up to weeks). Training AI on wind, weather and turbine information allows it to exceed the skills of traditional statistical approaches. Any forecast that accurately shows weather patterns will allow grid operators to dispatch electricity more efficiently, reduce fossil-fuel backup and prevent as much power loss. Being able to forecast helps in energy trading, where finding future prices and supplies from energy markets matters (Kılkış 2021). If we can depend on our predictions, it allows us to avoid maintenance in busy seasons and avoid missing out on key sales. LSTM networks, gradient boosting and mixes of CNN-LSTM are particularly skillful in discovering wind's unusual changes. It is possible to use storage more effectively when surplus or deficits can be predicted in advance. The information provided also helps control wind turbines through predicting upcoming events. Having predictive capability makes the system quicker, safer and more profitable. As we keep learning, the forecast model becomes more stable over time. Accurate weather forecasting is vital for both operating safely and economically in wind energy power systems.

5.3.7. *Enhanced turbine performance*

Improved turbine performance is possible when AI is used to observe, understand and control turbine operations in real time. The goal is to operate the turbine at its highest output while applying light mechanical pressure on it and wasting little energy. AI algorithms examine operational values and help make changes to how the blades tilt, point or operate via the generator. The statistics for performance metrics are analyzed to spot poorly performing turbines. AI suggests or does control actions that help run smoother and become more effective (Di Piazza et al. 2020). Here, predictive maintenance helps reduce periods of disruption and stop damage from occurring. AI contributes to better performance of turbines that is responsive to changes in wind and the electrical grid. Optimization in real time helps the turbines get as much output as possible. Making decisions in real time decreases the need for people to take action at every step and helps things move smoothly. Because it constantly learns from performance trends, AI can gradually enhance how turbines are run. Better performance helps equipment last longer, generate more energy annually and give us a better return on what we invested. It makes renewable energy a more trusted part of the energy system. AI, in short, greatly increases how efficiently turbines run compared to traditional manual control.

5.3.8. *Reliable grid management*

Stable grid management allows the grid to run properly, especially because the input from wind power can vary. AI makes it possible by arranging supply and demand, finding unusual situations and responding to them instantly. To manage a grid, voltage control, frequent changes and load coordination must be handled and this can now be done automatically by AI systems. Grid demand, forecasts and the output from wind power are all examined by these systems to choose immediate actions. Such tools are used by operators to forecast potential stress on the grid and make suitable plans (Elmousalami 2020). Demand response and dynamic pricing are carried out using both smart grid technologies and AI. AI, used with energy storage systems, helps to set the best times for charging and discharging. As a result, we can make use of the extra energy generated during high-wind events at any time. In addition, AI technology helps orchestrate energy sharing between different energy sources and the main grid. Stronger grid management can help suppress blackouts and improve the quality of electricity. With more renewable sources connected to the grid, it becomes more important to manage it dependably. The use of AI helps wind energy enhance the strength and sustainability of our electrical network.

5.4. Results and discussion

Figure 5.2 shows the variation in wind speed for a 24 hour period. Figure 5.3 shows the wind turbine power output over 24 hours. The power output peaks during early morning (5–8 am), and wind speed will be high during this time. In the afternoon, it decreases and begins to recover slightly in the evening. Figure 5.4 compares the actual and forecasted wind power output from AI model for a 24 hour period.

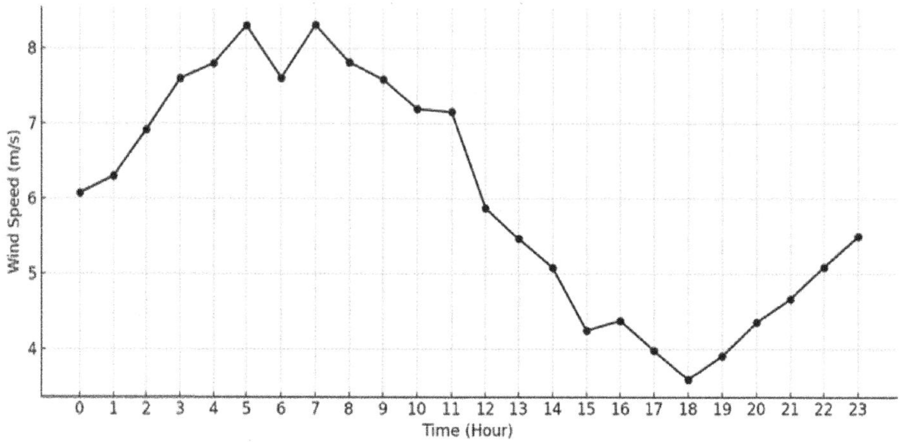

Figure 5.2. *Wind speed variation with time for 24 hours*

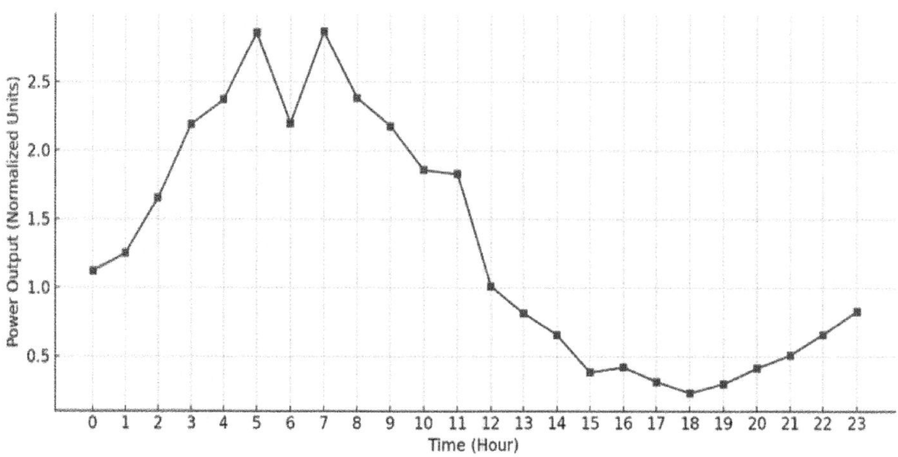

Figure 5.3. *Turbine power generation variation with time for 24 hours based on wind speed*

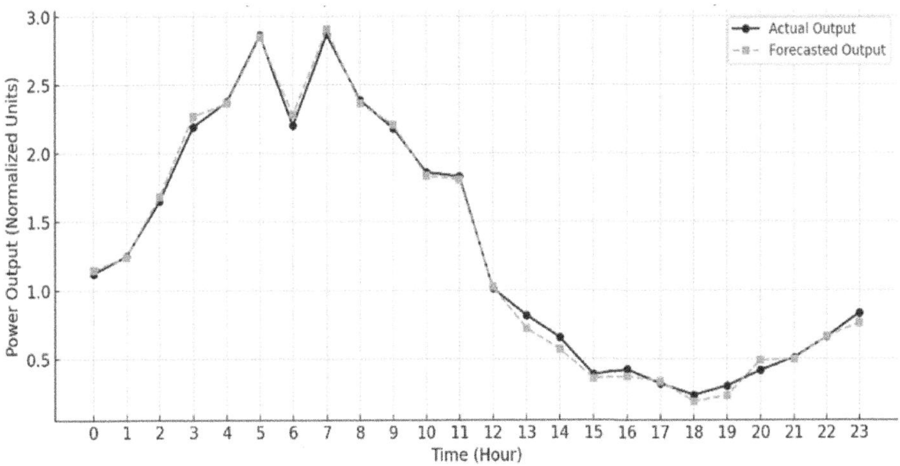

Figure 5.4. *AI-based power output with actual output from wind turbine for 24 hours. For a color version of this figure, see www.iste.co.uk/kumar/windpower2.zip*

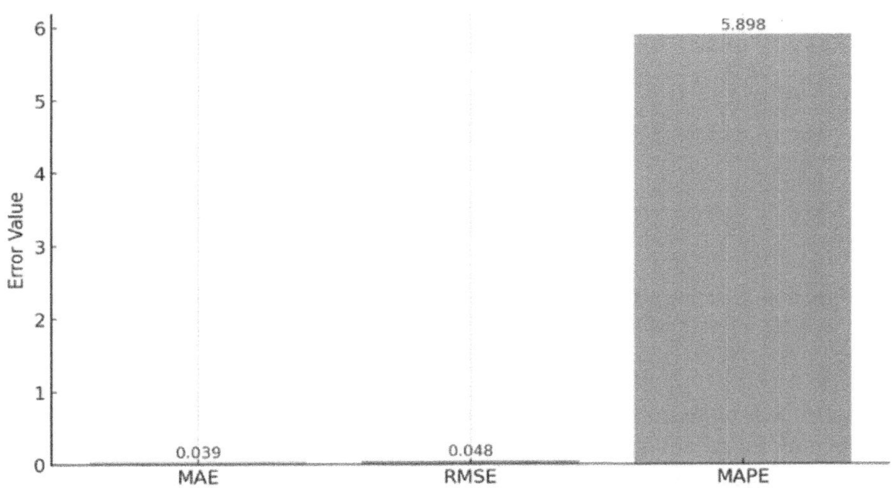

Figure 5.5. *Model error metrics comparison. For a color version of this figure, see www.iste.co.uk/kumar/windpower2.zip*

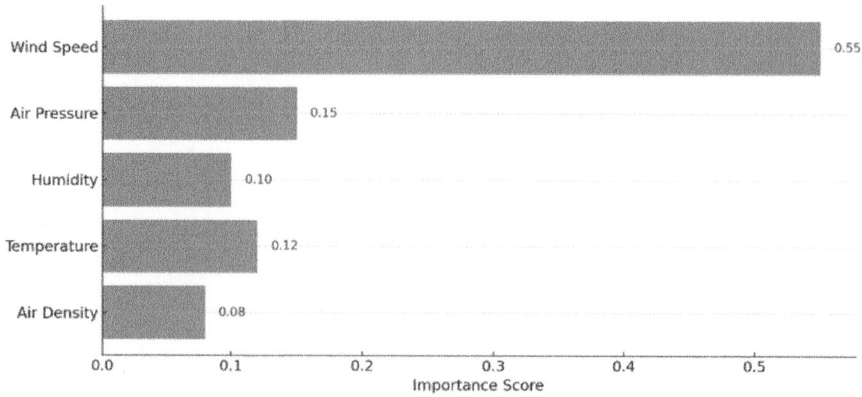

Figure 5.6. *Importance score for wind power forecasting*

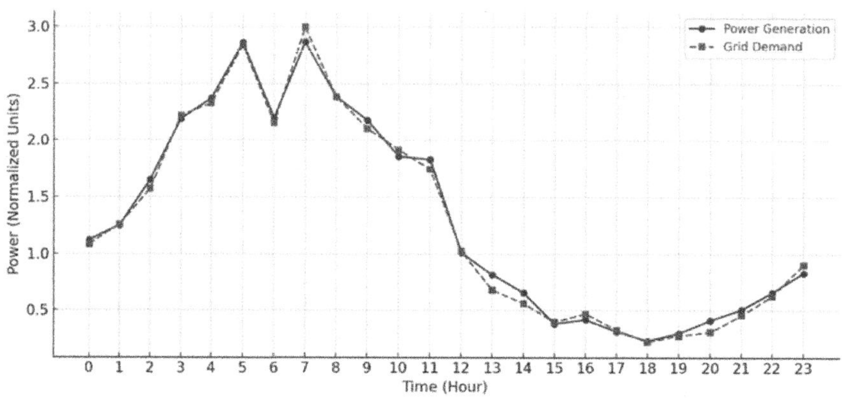

Figure 5.7. *Comparison of energy generated by a wind turbine system. For a color version of this figure, see www.iste.co.uk/kumar/windpower2.zip*

Figure 5.5 represents the model performance using three error metrics and reflects forecast reliability for wind power prediction. Figure 5.6 ranks the influence of input variables on wind power prediction and the wind speed with a score of 0.55.

Figure 5.7 compares the power generation and grid demand for a 24 hour period, which improves forecasting or energy storage integration.

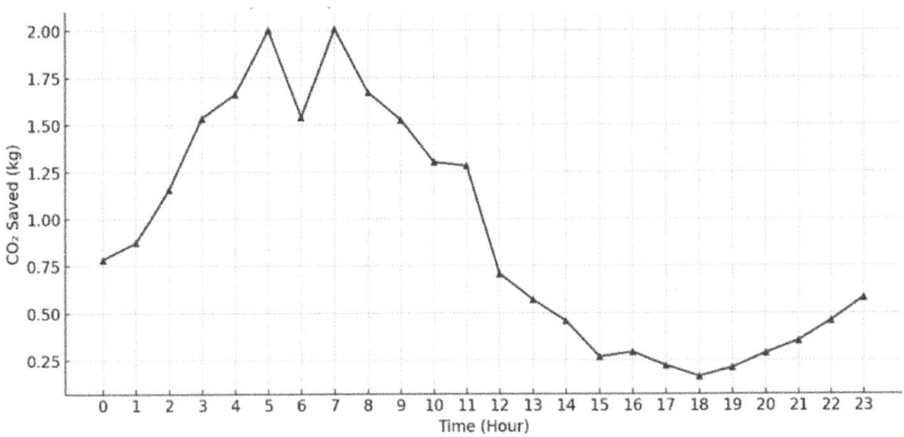

Figure 5.8. *CO_2 emission reduction over time*

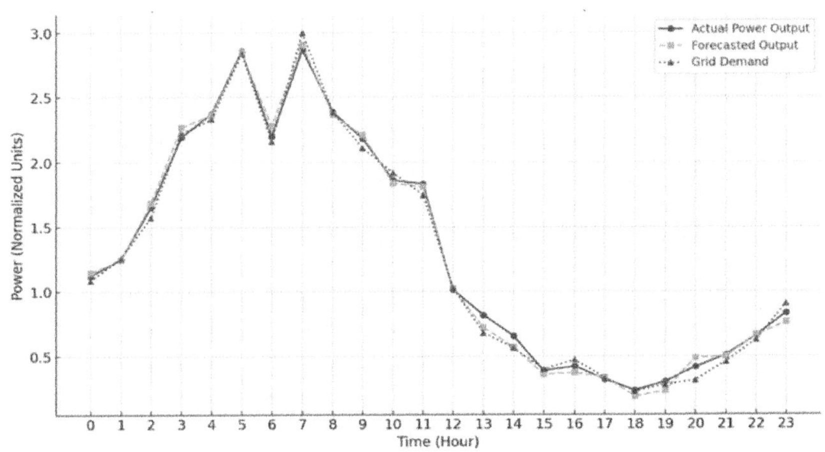

Figure 5.9. *Comparison chart for actual power output, forecasted output and grid demand. For a color version of this figure, see www.iste.co.uk/kumar/windpower2.zip*

Figure 5.8 gives the CO_2 emissions saved for a 24 hour period through wind energy generation, and Figure 5.9 represents the comparison chart for actual power output, forecasted output and grid demand for 24 hours, which indicates high forecasting accuracy and effective synchronization with energy demand.

Hour	Actual power output	Forecasted output	Grid demand	Forecast error	Demand gap
0	1.072	1.083	0.985	0.011	0.087
5	2.503	2.476	2.543	0.027	-0.04
10	1.869	1.907	1.893	0.038	-0.024
15	0.42	0.457	0.384	0.037	0.036
20	0.339	0.342	0.171	0.002	0.169
23	0.848	0.888	0.849	0.04	-0.001

Table 5.1. *Comparison table for the actual power output, forecasted values, grid demand, forecast error and demand gap*

Table 5.1 represents the comparison table for the actual power output, forecasted values, grid demand, forecast error and demand gap in which the forecasting accuracy is high and demand gaps fluctuate slightly with over or under generation.

5.5. Conclusion and future work

Technology in wind power makes the entire process of creating energy more efficient, dependable and sustainable. Using machine learning, the system can accurately predict wind data and produce an energy output that meets demand from the main grid. AI in turbine control helps achieve better performance and saves machinery parts from undue wear. Such strategies identify likely problems in advance, which reduces expenses and makes the turbines available more often. When linked to smart grids, energy can be moved efficiently to avoid overload and support energy demand. The simulations demonstrate that AI can produce forecasts with little error and remain near to the actual curve for both power and demand. It is determined that wind speed is the factor that provides the greatest influence on the accuracy of output predictions. The method decreases the reliance on fossil fuels and brings down carbon emissions. Such an energy network is making it possible for countries to turn to cleaner, more intelligent ways of producing energy. It shows how AI will help in producing cleaner energy.

Future research will pay attention to making AI models operate correctly and efficiently under different weather and location conditions. The use of hybrid deep learning and reinforcement learning in turbine control may result in better flexibility. Putting these tools into action in huge wind farms and connecting them with Internet of Things (IoT) devices for data collection is another good direction. Boosting effectiveness of energy storage is important so that surplus wind energy can be stored and used on days when wind is slow. Reducing the environmental effect of training and simulation in AI should be accomplished using Green AI approaches. It is necessary for energy regulatory and grid management policies to match with intelligent energy systems for a good and orderly adoption process. Giving operators and stakeholders user-friendly dashboards with decision tools will help them. In general, the advancement of AI for wind energy will come from building it to be more independent, powerful and conservation-oriented.

5.6. References

Alnaser, A.A., Maxi, M., Elmousalami, H. (2024). AI-powered digital twins and Internet of Things for smart cities and sustainable building environment. *Applied Sciences*, 14, 12056.

Cakiroglu, C., Demir, S., Hakan Ozdemir, M., Latif Aylak, B., Sariisik, G., Abualigah, L. (2024). Data-driven inter pretable ensemble learning methods for the prediction of wind turbine power incorporating SHAP analysis. *Expert Syst. Appl.*, 237, 121464. doi: 10.1016/j.eswa.2023.121464.

Cruz, L., Gutierrez, X.F., Martínez-Fernández, S. (2024). Innovating for tomorrow: The convergence of SE and Green AI. *arXiv Preprint*. doi: 10.48550/arXiv.2406.18142.

Dhayal, K.S., Giri, A.K., Agrawal, R., Agrawal, S., Samadhiya, A., Kumar, A. (2025). Do the innovative technological advancements foster the green transition pathways for industry 5.0: A perspective toward carbon neutrality. *Benchmarking*, 33(4), 1070–1098. doi: 10.1108/BIJ-04-2024-0330.

Di Piazza, A., Di Piazza, M., La Tona, G., Luna, M. (2020). An artificial neural network-based forecasting model of energy-related time series for electrical grid management. *Math Comput. Simul.*, 184, 294–305.

Elmousalami, H. (2020). Artificial intelligence and parametric construction cost estimate modeling: State-of the-art review. *J. Constr. Eng. Manage.*, 146, 03119008. doi: 10.1061/(ASCE)CO.1943-7862.0001678.

Elmousalami, H. and Sakr, I. (2024a). Artificial intelligence for drilling lost circulation: A systematic literature review. *Geoenergy Sci. Eng.*, 239, 212837.

Elmousalami, H. and Sakr, I. (2024b). Automated lost circulation severity classification and mitigation system using explainable Bayesian optimized ensemble learning algorithms. *J. Petrol. Explor. Prod. Technol.* doi: 10.1007/s13202-024-01841-4.

Elmousalami, H., Elmesalami, H.H., Maxi, M., Farid, A.A.M., Elshaboury, N.A.T. (2024). A comprehensive evaluation of machine learning and deep learning algorithms for wind speed and power prediction. *Decis. Anal. J.*, 3, 100527.

Elmousalami, H., Alnaser, A.A., Hui, F.K.P. (2025). Sustainable AI-driven wind energy forecasting: Advancing zero-carbon cities and environmental computation. *Artificial Intelligence Review*, 58, 191. doi: 10.1007/s10462-025-11191-0.

Fukutomi, M. (2024). Oil or geopolitical issues: Quantitative rethinking of political instability in the Middle East and North Africa. *GeoJournal*, 89, 55. doi: 10.1007/s10708-024-11045-2.

Karijadi, I., Chou, S.-Y., Dewabharata, A. (2023). Wind power forecasting based on hybrid CEEMDAN-EWT deep learning method. *Renew Energy*, 218, 119357.

Kılkış, S. (2021). Transition towards urban system integration and benchmarking of an urban area to accelerate mitigation towards net-zero targets. *Energy*, 236, 121394.

Liao, S., Tian, X., Liu, B., Liu, T., Su, H., Zhou, B. (2022). Short-term wind power prediction based on light GBM and meteorological reanalysis. *Energies*, 15, 6287. doi: 10.3390/en15176287.

Malakouti, S.M. (2023). Estimating the output power and wind speed with ML methods: A case study in Texas. *Case Stud. Chem. Environ. Eng.*, 7, 100324.

Malakouti, S.M. and Ghiasi, A.R. (2022). Evaluation of the application of computational model machine learning methods to simulate wind speed in predicting the production capacity of the Swiss Basel wind farm. In *26th International Electrical Power Distribution Conference (EPDC)*. IEEE, 31–36 [Online]. Available at: https://ieeexplore.ieee.org/abstract/document/9817304/.

Malakouti, S.M., Ghiasi, A.R., Ghavifekr, A.A., Emami, P. (2022). Predicting wind power generation using machine learning and CNN-LSTM approaches. *Wind Eng.*, 46, 1853–1869. doi: 10.1177/0309524X22 1113013.

Ramaswami, A., Tong, K., Canadell, J.G., Jackson, R.B., Stokes, E., Dhakal, S., Finch, M., Jittrapirom, P., Singh, N., Yamagata, Y. et al. (2021). Carbon analytics for net-zero emissions sustainable cities. *Nat. Sustain.*, 4, 460–463. doi: 10.1038/s41893-021-00715-5.

Seto, K.C., Churkina, G., Hsu, A., Keller, M., Newman, PWG., Qin, B., Ramaswami, A. (2021). From low- to net-zero carbon cities: The next global agenda. *Annu. Rev. Environ. Resour.*, 46, 377–415. doi: 10.1146/annurev-environ-050120-113117.

Tornede, T., Tornede, A., Hanselle, J., Mohr, F., Wever, M., Hüllermeier, E. (2023). Towards green automated machine learning: Status quo and future directions. *J. Artif. Intell. Res.*, 77, 427–457.

Van Wynsberghe, A. (2021). Sustainable AI: AI for sustainability and the sustainability of AI. *AI Ethics*, 1(213), 218. doi: 10.1007/s43681-021-00043-6.

Verdecchia, R., Sallou, J., Cruz, L. (2023). A systematic review of Green AI. *Wires Data Min. Knowl.*, 13, e1507. doi: 10.1002/widm.1507.

Wang, C., Song, J., Shi, D., Reyna, J.L., Horsey, H., Feron, S., Zhou, Y., Ouyang, Z., Li, Y., Jackson, R.B. (2023). Impacts of climate change, population growth, and power sector decarbonization on urban building energy use. *Nat. Commun.*, 14, 6434. doi: 10.1038/s41467-023-41458-5.

Wilson, C. and Van Der Velden, M. (2022). Sustainable AI: An integrated model to guide public sector decision making. *Technol. Soc.*, 68, 101926.

Wu, C.-J., Raghavendra, R., Gupta, U., Acun, B., Ardalani, N., Maeng, K., Chang, G., Aga, F., Huang, J., Bai, C. (2022a). Sustainable AI: Environmental implications, challenges and opportunities. *Proc. Mach. Learn. Syst.*, 4, 795–813.

Wu, Z., Luo, G., Yang, Z., Guo, Y., Li, K., Xue, Y. (2022b). A comprehensive review on deep learning approaches in wind forecasting applications. *CAAI Trans. Intel. Tech.*, 7, 129–143. doi: 10.1049/cit2.12076.

Yaghoubirad, M., Azizi, N., Farajollahi, M., Ahmadi, A. (2023). Deep learning-based multistep ahead wind speed and power generation forecasting using direct method. *Energy Convers. Manage.*, 281, 116760.

6

Economic and Market Impacts of AI in Wind Power

The application of Artificial Intelligence (AI) in wind energy development is revolutionizing the economic and market dynamics of the renewable energy sector. In this chapter, the impact of AI-driven innovations on cost structures, investment trends, market competitiveness and operational efficiencies of the wind energy industry is critically examined. The research uses a mixed-methods design, integrating quantitative economic modeling with qualitative market trend analysis. Global and regional wind energy market data are examined to determine the effect of AI applications such as predictive maintenance, dynamic pricing and energy trading algorithms on capital and operational costs. Research evidence suggests that AI use reduces maintenance costs to 30%, increases turbine uptime and enhances energy yield forecasting accuracy, leading to increased financial returns. Market studies point towards a correlation between the usage of AI and increased investor confidence, as well as accelerating commercialization of wind technologies. The chapter reaches the conclusion that AI is an essential driver of economic sustainability and market expansion for the wind power industry. Targeted use of AI technologies will reduce the levelized cost of energy (LCOE) and develop novel business models, thus further anchoring wind power in the universal shift towards renewable energy.

6.1. Introduction

With the world grappling with mounting environment and energy crises, the shift to clean energy has become one of our greatest priorities in this generation. Global warming, fueled primarily by carbon dioxide emissions from fossil fuels, continues to wreak havoc on the environment and imperil economic stability worldwide (Bošnjaković et al. 2025). National governments, global bodies and private

Chapter written by Mantena Siva Pavan Kumar RAJU and Mantena SIREESHA.

enterprises have therefore stepped-up efforts to move towards cleaner energy. Among the renewable portfolio, wind energy has emerged as one of the most developed, scalable and economic technologies (Sun and Han 2025). As per the Global Wind Energy Council (GWEC), global capacity for wind power crossed 900 GW in 2023, highlighting a robust growth path. Despite this momentum, the wind industry remains plagued by determinative constraints, most prominently those related to high capital cost, the variability of power output, losses in efficiency during operation and the challenge of incorporating wind electricity into increasingly complex and liberalized power markets (Mitchell et al. 2022; Harrison-Atlas et al. 2024). These constraints prevent the full potential of wind power from being realized and emphasize the need for revolutionary technological solutions.

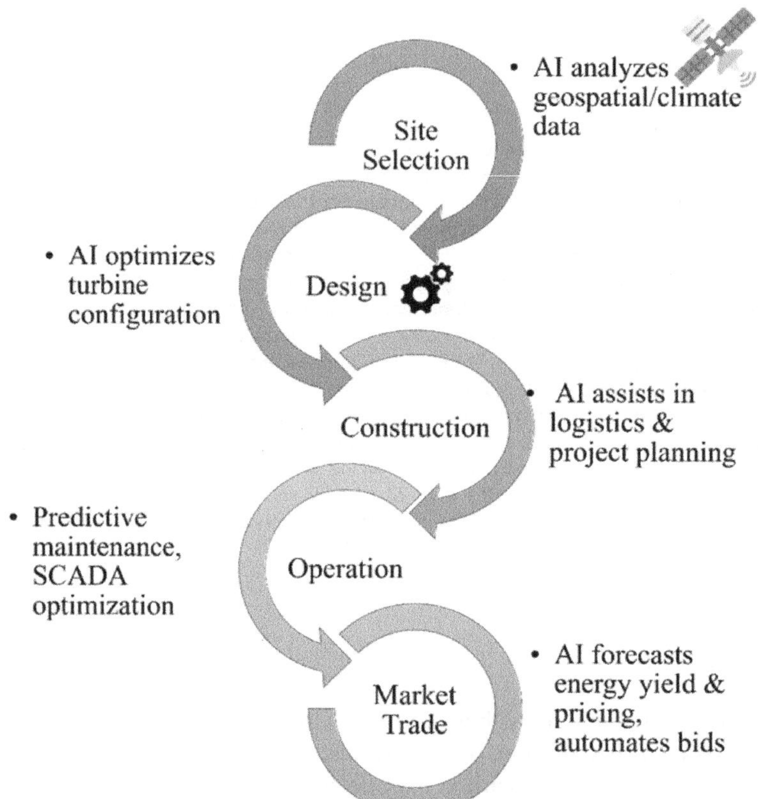

Figure 6.1. *AI applications across wind energy lifecycle. For a color version of this figure, see www.iste.co.uk/kumar/windpower2.zip*

Here, AI has shown itself to be a promising field that can help solve several pain points in the wind energy value chain (Ohalete et al. 2023). AI technologies such as machine learning, predictive analytics and intelligent automation have already started to re-engineer renewable energy business operations (Mauludin et al. 2024). In wind energy in particular, AI is used to improve wind forecasts, improve turbine efficiency, automate maintenance planning and enhance trading plans for energy (Dehshiri and Firoozabadi 2024). Geospatial data, land formations and climate conditions at planning and construction times are read by AI software in order to dictate locations for installations of turbines (Lipu et al. 2021). Once in use mode, the real-time data of the SCADA system along with IoT sensors inform AI-based systems to predict mechanical breakdowns, recommend optimum performance and alter blade pitch and yaw position to cope with the variability in the winds (Mitchell et al. 2022). These uses not only enhance turbine efficiency but also minimize downtime, increase asset lifetimes and ultimately enhance the financial returns of investors and developers.

AI offers tangible advantages in discrete aspects of wind energy system operations as depicted in Figure 6.1. For instance, AI-based predictive maintenance has demonstrated the ability to save maintenance costs up to 30% and cut unplanned outages to a minimum, contributing to higher turbine availability (Acaroğlu and Márquez 2021). Similarly, AI-enabled forecasting solutions have increased the accuracy of wind power forecasts, reducing forecasting errors from 20% to below 5%, hence enabling more accurate bidding in electricity markets (Al-Othman et al. 2021). However, despite these individual technological milestones being better recognized, still missing is the greater understanding of the overall economic and market-level implications of AI adoption (Chatterjee and Dethlefs 2021b). Most of the current literature is technologically focused on performance parameters and pays less attention to how the adoption of AI influences overall capital expenditure (CAPEX), operational expenditure (OPEX), levelized cost of energy (LCOE), return on investment (ROI) and competitiveness of the business in the marketplace. In addition, local variations in AI take-up, investment preparedness and policy frameworks are underemphasized despite their significance to the scaling up of wind power in various contexts around the world (Jiang and Liu 2021). There is also limited understanding of the use of AI to influence future business models and structures of energy markets. Upcoming trends, including the application of AI for dynamic pricing, peer-to-peer energy trading and AI-as-a-Service (AIaaS) platforms, indicate that AI has the potential to transform the financing, marketing and consumption of wind energy fundamentally. Yet, the economic rationale and long-term sustainability of these developments remain in preliminary stages of evaluation (Msigwa et al. 2022). There is no literature linking the technical strengths of AI with macroeconomic indicators such as patterns of investment, risk mitigation and policy responses. Such a knowledge gap is especially essential now that the international wind industry is trending towards higher decentralization and digitalization where strategic AI deployment will define not only the operational effectiveness but also the financial viability and scalability. Against this, the aim of this chapter is to analyze the economic and market impacts of AI in the wind power sector in a cost-efficient and market-sensitive manner.

This encompasses an in-depth discussion on how AI influences economic performance indicators such as CAPEX, OPEX, ROI and LCOE and how it intensifies competition within electricity markets with improved forecasting, automated bidding and dynamic pricing models. The chapter also aims to examine the impact of AI on investor behavior, funding models and new market structures such as peer-to-peer energy trading and performance-based power purchase agreements (PPAs). Through leveraging industry data, case studies and economic modeling, the research aims to develop a complete picture of AI as a strategic enabler – not only of technical performance, but also of economic change in wind energy. The hypothesis that underpins this research is that increased integration of AI technologies across the wind energy value chain generates measurable improvements in economic viability and market adaptability. In particular, the chapter argues that AI lowers LCOE by maximizing operational workflows, improves ROI through increased turbine uptime and revenue consistency and shortens the payback period via cost savings. Also, it is theorized that AI promotes investor confidence by making cash flows more predictable and allowing for smarter, data-driven decision-making. These benefits, when achieved, will place AI not as a marginal technological enhancement, but as a fundamental component in the strategic design and deployment of wind energy systems worldwide. Finally, this chapter hopes to fill the existing knowledge gap between technical advancements and economic performance in AI-powered wind energy systems. It presents a vision-oriented perspective of how AI can not only be used to overcome existing challenges but to create new opportunities across finance, policy and market engagement. In doing so, it joins the expanding pool of interdisciplinary scholarship examining digitalization as a driver of sustainable, scalable and economically viable energy solutions amidst global climate and energy imperatives.

6.2. Operational efficiencies and cost savings through AI

6.2.1. *Predictive maintenance and reduced downtime*

The most innovative application of AI in the wind power industry is predictive maintenance, which significantly lowers unplanned turbine shutdowns and extends the service life of critical components (Chen et al. 2021b). The classical maintenance model traditionally relied significantly on intermittent inspection or repair work upon breakdown, both of which were inefficient regarding availability of resources, safety and cost (Onwusinkwue et al. 2024). Unlike this, AI-based systems monitor turbine operating conditions continuously in real time, such as vibration patterns, blade stress levels, temperature changes and acoustic signatures, to detect any symptoms of wear or failure at the onset (Elmousalami et al. 2025). Based on sophisticated machine learning algorithms, these systems can predict probable failures before they happen, and maintenance teams can be ready ahead of time (Shahid et al. 2024). This paradigm shift results in staggering cost savings in maintenance estimates of as much as 30%, which have been substantiated through research and a balancing increase in turbine availability (Thapa 2022). Lower downtime not only enhances the efficiency of operation but also stabilizes energy output, having a direct bearing on financial predictability, and enhances investor confidence (Rusilowati et al. 2024).

6.2.2. *Turbine performance enhancement through AI*

Above maintenance, AI technologies play a key role in maximizing turbine performance during energy production as shown in Table 6.1. AI systems combine Supervisory Control and Data Acquisition (SCADA) inputs, sensor arrays and historical performance records to adjust turbine operations dynamically according to changing wind conditions (Loi et al. 2023). Algorithms are applied to optimize blade pitch angles, yaw alignment and generator settings for maximum aerodynamic efficiency (Thapa 2022). This real-time adaptive control is particularly valuable in sites where the wind patterns change, harvesting more energy with lower structural stresses and fatigue (Machado and De Andrés 2022). AI can also feel micro-level inefficiencies, such as wake effects between turbines at big farms, and change configurations to reduce power losses (Chatterjee and Dethlefs 2021b). Intelligent operations such as these lead to an overall boost in energy yield by 3–5% above traditional systems, without extra physical infrastructure investment.

Metric	Traditional wind farm	AI-optimized wind farm
OPEX ($/MWh)	20–30	12–18
Downtime (hrs/year/turbine)	150–200	50–70
Energy yield accuracy (%)	±10%	±2–3%
Forecasting error (%)	~20%	<5%

Table 6.1. *Traditional versus AI-optimized wind farm metrics. Sources: IRENA Wind Technology Fact Sheets; GE Renewable Energy AI maintenance reports*

6.2.3. *Effect on operating and capital spending*

Efficiencies brought about by AI have a direct correlation with quantifiable reductions in both OPEX and, to some degree, CAPEX. In terms of OPEX, cost savings are achieved through fewer maintenance interventions, less labor demand, less downtime and better energy dispatch optimization (Stecyk and Miciuła 2023). For example, AI-optimized wind farms have OPEX ranging from $12 to 18/MWh compared to $20–30/MWh for conventional farms. Additionally, predictive analytics lower the rate of expensive emergency overhauls and increase component longevity, thus decreasing long-term operation expenses (Onwusinkwue et al. 2024). In the case of CAPEX, though the incorporation of AI may drive the initial cost of deployment higher with costly sensors, data platforms and analysis tools initially, long-term economic return usually overcomes the initial expense (Chen et al. 2021b). By leveraging AI-driven environmental and climatic research for advanced site selection, developers can avoid slum areas and reduce high development costs (Machado and De Andrés 2022).

Optimization of intelligent design also allows for leaner, more efficient turbine structures with maximum structure and space efficiency (Barbosa et al. 2023). These common reductions in OPEX and CAPEX are among the reduction of LCOE, a significant metric of economic feasibility of wind projects (Rusilowati et al. 2024). With the use of AI, LCOE decreases due to higher capacity factors, improved forecasting and improved asset utilization (Stecyk and Miciuła 2023). Figure 6.2 displays a noticeable inverse relationship between the levels of AI adoption and LCOE and also reflects the economic justification of digitalization for wind. In addition to this, AI also enhances compliance with grid codes and markets by reducing cost fines and enabling closer interaction with trading platforms for energy (Wang et al. 2023). By maximizing the technical and financial dimensions of wind power generation, AI is an effective driver of renewable deployment on a global basis.

6.3. Influence of AI on market dynamics

Not only are AI technologies revolutionizing the operational dynamics of wind power projects, they are also completely altering the energy market dynamics. With improved accuracy in predictions, enabling real-time engagement in the market and facilitating dynamic pricing mechanisms, AI renders wind energy generators more competitive and responsive to advanced electricity markets (Elmousalami et al. 2025). These capabilities are particularly vital in deregulated and decentralized power networks, where responsiveness to markets and predictive acumen are the secrets to profitability and grid stability.

6.3.1. *AI in energy forecasting and energy trading*

Perhaps the most important contribution of AI in market operations is its application in energy forecasting. Wind power, being variable and weather-dependent, poses a consistent challenge in predicting power generation accurately (Shahid et al. 2024). Most traditional forecasting techniques have difficulty in coping with the randomness of wind and produce high forecasting errors often in the order of 20%, thus leaving energy producers vulnerable to imbalance penalties and revenue instability (Khalid 2024). AI algorithms, particularly machine learning and deep learning approaches, are very significant improvements (Barbosa et al. 2023). By ingesting massive quantities of data, historical weather trends and live meteorological feeds to turbine operation-specific data AI systems can make predictions with wildly reduced error rates, sometimes less than 5% (Elmousalami et al. 2025). This improved predictive ability enables wind farm operators to optimize power delivery scheduling, prevent monetary losses in the form of penalties from grid operators and participate in more lucrative market bidding strategies (Onwusinkwue et al. 2024).

Enhanced forecasting makes effective participation in energy trading markets possible. AI-powered trading platforms can make automated bids, price reallocations and hedging strategies on the basis of real-time market indicators and predictive models (Yousef et al. 2023). Such platforms dynamically evaluate supply–demand situations, competitor activity and grid congestion levels to take data-driven trading decisions (Zhao et al. 2022). For instance, if wind generation is forecasted to decline, AI platforms can proactively reprice or explore other trading alternatives, thus avoiding revenue loss. The ability to automate and optimize trading in such highly grained and reactive manners is an unequivocal competitive advantage for AI-integrated wind farms.

6.3.2. *Dynamic pricing models and market participation*

AI also enables the development and implementation of dynamic pricing regimes, which replace static tariffs with more flexible, real-time price regimes (Khan and Nasir 2023). These frameworks tie energy prices to demand, availability and market variation. AI-driven wind power turbines can calculate when best to deliver electricity based on predicted prices, harvesting the highest income during high-demand hours and keeping losses at bay during low-price hours. In addition, dynamic models help in the overall market efficiency by more effectively matching generation to consumption (Thapa 2022). In peer-to-peer trading energy markets, AI can facilitate localized energy trading as well as peer-to-peer trading platforms and direct trading between individual producers and consumers (Zhao et al. 2024). Blockchain technology-based smart contracts, supported by AI-verification and settlement platforms, facilitate transparency, secure and automated trade. Such technologies are particularly beneficial for community wind projects and microgrids to enable small-scale producers to significantly contribute to the energy economy.

6.3.3. *AI-assisted wind farms: case studies*

A few instances from the real world show the disruptive market impact of AI implementation in wind energy (Franki et al. 2023). For example, GE Renewable Energy has applied advanced AI solutions to its wind turbines to enable real-time diagnostics, predictive maintenance and performance optimization, which yields up to a 20% boost in accuracy in energy forecasting. Vestas has also leveraged AI algorithms to track turbine performance and optimize operations in real time to market signals, thereby boosting uptime and earnings. Such AI solutions are being used extensively in greenfield deployments and retrofitted for brownfield assets, a testament to their extended value proposition. In addition, pilot schemes in regions such as Europe and North America have demonstrated that wind farms that use AI perform better in market responsiveness, reduction in curtailment and ROI. In liberalized markets, these projects have proved to be more competitive in terms of their capacity to optimize bid scheduling and price in day-ahead and real-time markets. The experience of these pioneers offers a compelling model for future wind power developments, particularly in emerging economies seeking to build stable and cost-effective renewable infrastructure. Through empowering wind power generators with tools for predicting market trends, setting optimum prices and trading effectively, AI helps play the role of a force multiplier in implementing a transition towards smarter, reactive and stronger energy systems. Through its ability to turn variability into predictability and complexity into opportunity, AI has emerged as a key device to enable wind power operators to operate the fast-evolving and globally connected markets for energy.

6.4. Economic analysis of AI integration in wind projects

AI integration in wind power infrastructure has provided new grounds for assessing the economic viability and profitability of renewable energy projects (Kamisetty 2022). As much as the technological strengths of AI such as enhanced maintenance and performance enhancement are well recognized, its economic impacts require closer examination through detailed cost–benefit analysis, enhanced financial performance metrics and quantifiable impacts on LCOE (Al-Othman et al. 2021). The paradigms of integral economics guide the examination in this section to how AI redefines investment sense, optimizes financial returns and returns value for developers, investors and grid operators simultaneously.

6.4.1. *Cost–benefit analysis framework*

The right cost–benefit analysis of wind proposals with the use of AI entails reference to the initial investment in terms of CAPEX relative to the application of new technology, as well as long-term financial returns based on operational efficiency and revenue growth (Behara and Saha 2022). On the spend side, the use of AI comprises expenditure on digital infrastructure in the form of IoT sensors, data analytics solutions and software subscriptions (Mhlanga 2023). Also, training staff and integrating data with existing systems could bring transitional costs. However, these expenses are generally front-end and decrease over time as systems mature and grow.

On the benefit side, the economic gains are extensive. AI lowers unplanned maintenance, maximizes turbine efficiency, reduces insurance costs through improved risk profiling and reduces curtailment through improved forecasting (Rusilowati et al. 2024). For example, predictive maintenance alone can lower maintenance expenses by as much as 30%, and AI-based forecasting has been found to lower financial penalties for grid mismatches (Onwusinkwue et al. 2024). When these advantages are tallied over the lifetime of a project, typically 20–25 years, the internal rate of return (IRR) significantly improves. This activity creates the economic justification for deploying AI and merits the comparatively reasonable initial investment outlays.

6.4.2. *Return on investment (ROI) and payback period improvements*

ROI and payback period are two critical metrics relied upon by developers as well as financiers to determine project viability. Both of these are significantly enhanced by AI-driven wind projects. Increased turbine availability translates directly to additional AEP each year, increasing revenues (Yousef et al. 2023). At the same time, lower O&M costs translate directly to higher net profit margins. Case studies suggest possible 8–15% ROI enhancement over traditional installations with AI-compliant wind farms.

In addition, the payback period measured as the duration to return the initial investment decreases significantly in AI-based projects (Rane et al. 2023). Traditional wind farms take 7–10 years to break even depending on native wind resources and the cost of electricity. With AI efficiencies, this can be reduced by one to three years. This shorter payback period is especially appealing to investors who work within volatile or emerging markets, where quick financial agility and faster capital recovery are crucial to sustaining long-term energy growth (Msigwa et al. 2022).

6.4.3. *Effect on levelized cost of energy (LCOE)*

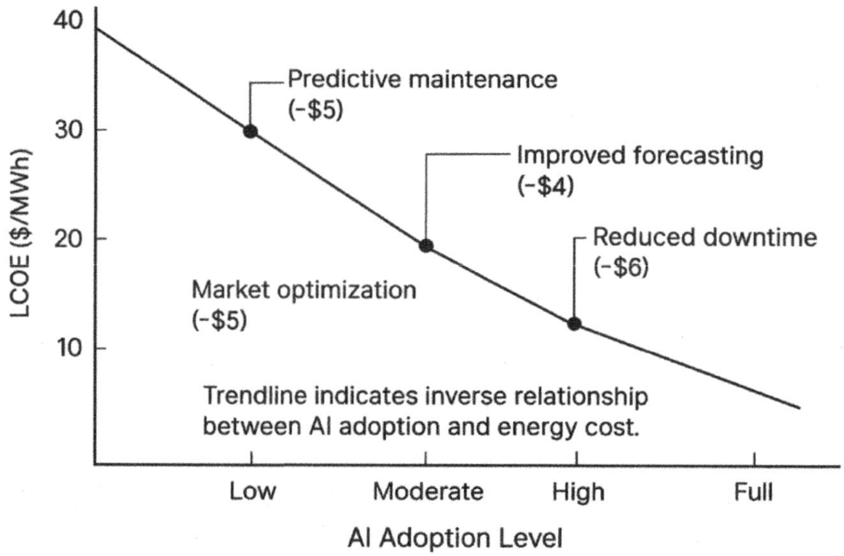

Figure 6.2. *Graph showing reduction in LCOE with increasing levels of AI adoption (hypothetical or real-world data trends). Source: Lazard 2023 Levelized Cost of Energy v17.0; IRENA 2023 Innovation Outlook on AI & Big Data; IEA 2024 Digital Demand-Driven Electricity Networks Roadmap; GE Renewable Energy 2024 AI-Enabled Predictive Maintenance White Paper; DNV 2024 Energy Transition Outlook–Digital Twins & Autonomous Turbines; BloombergNEF 2024 Digitalisation & AI in Renewables Market Outlook; GWEC 2024 Global Wind Report*

One of the strongest economic cases for integrating AI is found in its influence on LCOE, an indicator that bundles all lifetime cost and energy outputs into one figure represented in terms of $/MWh. LCOE is a key parameter of energy policy, investment and technology comparison. Both numerator (cost) and denominator (output) aspects of the LCOE formula may be enhanced by AI, thus offering the potential for a multidimensional assessment (Stecyk and Miciuła 2023). For instance, decreases in OPEX (displayed in Table 6.1) from \$20–30/MWh in conventional configurations to \$12–18/MWh in AI-optimized systems meaningfully reduce the cost factor. Concurrently, greater energy yield forecasting precision and lower downtime lead to higher overall energy output over the lifespan of the asset, thus enhancing the output parameter. In AI-integrated wind farms, LCOE savings of 10–25% have been achieved, depending on deployment scale and local conditions.

This is a revolutionary step, making AI not only an add-on feature but a central enabler of economic competitiveness. With hardware costs decreasing leveling out across the wind sector, software-based efficiencies such as AI are the next horizon for cost savings. Essentially, the economic value of AI is its ability to transform wind power projects from marginal to profitable, from risky to secure, and from fixed to flexible. By enhancing the financial architecture of projects, AI frees up capital, draws in private and institutional investment, and makes wind power a more bankable and scalable part of the world's energy mix.

6.5. Role of AI in wind power financing and investment trends

The role played by AI within wind power financing and investment processes is being felt ever more strongly. With its promise of increasing transparency, operational foreseeability and fiscal performance, AI is quickly emerging as one enabler among key drivers to spur confidence-based investments (Kamisetty 2022). In this section, we discuss how the adoption of AI alters investor patterns as in Table 6.2, explain various financing options and introduce new economic platforms to the opportunities of creation in the wind industry.

Year	Global investment (\$ billion)	Notable AI developments
2018	2.5	Pilot AI in maintenance
2020	4.2	Early adoption in energy trading
2022	6.8	Scaled AI forecasting tools
2024 (est)	10.1	Widespread AI + IoT deployments

Table 6.2. *Global investment growth in AI-enabled renewable projects. Sources: BloombergNEF (BNEF); International Energy Agency (IEA); Statista: https://www.statista.com*

6.5.1. *How AI supports investor confidence*

Investor confidence in wind investment comes with the ability to reduce risks, forecast financial performance and stay under regulation. All these areas owe significant contributions from AI technologies. Through predictive analytics, AI causes mechanical failure, weather disturbances and grid instability being detected at an early stage (Khalid 2024). These elements reduce uncertainty and interruptions to operations, rendering cash flows bankable and predictable (Rane et al. 2023). Secondly, advanced predictive capability fueled by AI elevates forecasting in energy delivery and optimizes the bidding approach towards the market (Khalid 2024). Such outcomes lessen exposure to the risk of variability in prices as well as the imbalance penalty and make investors well-informed regarding expected returns.

More financial institutions are now using datasets and performance models powered by AI during the due diligence process (Márquez and Gonzalo 2021). The integration of AI into digital twins' virtual representations of wind turbines and entire farms allows investors to simulate different operating conditions and estimate long-term performance with high precision (Shahid et al. 2024). Therefore, AI implementation not only plays operational functions but also serves as a sign of credibility in the eyes of capital providers. In essence, wind farms installed with AI are lower-risk, higher-reward ventures compared to those that rely on the traditional system.

6.5.2. *Impact on policy frameworks and financial risk management*

The application of AI on a large scale in wind energy has prompted a rethinking of regulatory and policy frameworks that govern funding for renewable energy (Simankov et al. 2024). Governments and global institutions increasingly see the power of AI to catalyze cost-effectiveness, enhance grid integration and hasten clean energy shifts (Khan and Nasir 2023). Consequently, policy tools such as green bonds, subsidies and feed-in tariffs are being reformed to benefit AI-integrated schemes (Jiang and Liu 2021). Further, AI-facilitated compliance monitoring such as tracking emissions, equipment safety and operational performance eases reporting requirements for regulators and increases project transparency (Zhao et al. 2022).

From a risk management point of view, AI allows for refined tracking of performance metrics on a project. Its greatest value is to institutional investors and insurance companies with an acute requirement for accountability (Jiang and Liu 2021). Through its ability to mine operating data in real time, AI can detect exceptions, establish performance targets and condition models of financial risk (Rane et al. 2023). This enhances the probability of unwanted events taking place and allows for more effective pricing of insurance and financial products. It also supports performance-based financial products, such as variable loan conditions or interest rates tied to performance of a model that is gaining popularity under ESG investing.

6.5.3. *Emerging AI-driven business models*

Most potentially disruptive of these changes, however, is the advent of new business models that break with conventional energy project frameworks. A case in point is AI-as-a-Service (AIaaS), where AI platforms provide subscription-based offerings to wind farm operators for forecasting, maintenance planning and market optimization (Li 2023). The model reduces the cost of adopting AI by eliminating the huge upfront costs of software and makes even small-scale wind producers eligible for high-end analytics (Behara and Saha 2022). Another new model includes decentralized energy trading platforms, and AI and blockchain technologies are converging to facilitate peer-to-peer energy trading (Zhang et al. 2024). These platforms function autonomously, authenticate transactions and dynamically adjust prices based on supply and demand analytics. Wind generation producers, particularly community-scale and small-scale developers, are now free to join local energy economies without reliance on centralized grid operators (Zhao et al. 2024). AI is also facilitating performance-based PPAs, where energy consumers pay for energy delivered rather than tariffs. AI ensures accurate forecasting and delivery, reducing the difference between expected and realized energy supply (Mhlanga 2023). The model optimizes the value proposition for both consumers and producers, encouraging long-term contractual relationships based on open performance data.

In short, AI is not merely an efficient optimization device, it is a strategic enabler of financial innovation and investor engagement. Through reduced risk, optimized performance predictability and support for new economic frameworks, AI is reshaping the financing, operation and development of wind projects globally. With the progress of AI, its role in shaping the financial form of the wind power industry will deepen, with it becoming an integral pillar in the pursuit of a low-carbon economy.

6.6. Limitations and challenges

While integration of AI in the wind power sector is replete with remarkable operational and economic benefits, it is not without vulnerabilities (Chang et al. 2021). Implementation of AI technologies at various points of the wind energy value chain is plagued with grave challenges relating to expenses, data management, infrastructure supportability and preparedness of the workforce (Lipu et al. 2021). Addressing these weaknesses is necessary to provide assurance for the fair, secure and sustainable deployment of AI solutions to the world wind power market (Rhatrif et al. 2024). The next section presents the greatest challenges to the deployment of AI and how these can delay or prevent the realization of AI's full potential in wind energy.

6.6.1. *High upfront integration expenses*

One of the biggest hindrances to embracing AI is the prohibitive installation price tag. Embracing AI solutions typically incurs massive investments in hardware, software and data capturing devices (Nazir et al. 2022). This involves the installation of sophisticated IoT sensors, the replacement of SCADA systems, high-performance computing devices and cloud storage facilities (Mitchell et al. 2022). To most wind power developers, particularly in developing nations or small-scale installations, these upfront costs are unaffordable.

In addition, AI-based solutions must be tailored to every site, considering the model and manufacturer of the turbine, weather conditions and grid connection requirements (Machado and De Andrés 2022). These tailoring efforts are likely to require specialized engineering skills and long deployment schedules (Acaroğlu and Márquez 2021). The resulting cost drivers have the effect of deterring investors and developers from applying AI, particularly if returns are unreal or delayed (Chang et al. 2021). Even though economies of scale and open-source AI software will presumably reduce such expenses in the future, cost considerations today are a significant barrier to large-scale deployment.

6.6.2. *Data privacy and cybersecurity issues*

Yet another huge issue surrounding AI-driven wind systems is data privacy and security (Rhatrif et al. 2024). AI solutions implemented in wind farms heavily depend on the transmission of real-time data, remote monitoring and cloud-based or central processing (Yu et al. 2023). All this tremendous data exchange provides new avenues for cyberattacks that could be directed towards critical infrastructure, jeopardize operational safety or manipulate energy trading algorithms (Wang et al. 2023). Cybersecurity is even more intricate in multi-vendor settings with many components and platforms communicating with one another. Safe data exchange among various systems requires strong encryption algorithms, frequent vulnerability scanning and quick incident response systems (Sofian et al. 2024). Moreover, data privacy regulatory environments are changing and considerably varied across geographies, which creates compliance difficulties for worldwide developers.

6.6.3. *Skills gap in AI implementation*

Effective use of AI in wind energy also calls for skilled labor with proper combinations of domain knowledge and digital skills. The intersection of these skills – wind engineering, data science, machine learning and system integration – is highly sought after in most developing economies of the present day (Liu et al. 2022).

Without such a deficit, the market might be exposed to inappropriate deployment, waste of AI solutions and excessive learning for operations and maintenance staff. Additionally, AI systems also need periodic updates, algorithm retraining, and bias or prediction drift monitoring (Khan and Nasir 2023). These are complicated processes needing continuous attention from competent personnel. Without adequate training and supervision, AI systems can yield erroneous or biased results, leading to performance degradation or safety risks (Nazir et al. 2022). To bridge this shortfall, the industry players and the education system will have to implement specialized training programs in digital energy systems. Reskilling the current energy professionals and forming multidisciplinary teams will be required to leverage the potential of AI.

6.7. Future opportunities

With the advancement of the wind power industry, the synergy of AI and newer digital and decentralized technologies holds a vast potential to reshape the energy future of the world (Ukoba et al. 2024b). Development in one of the key areas in the future is through working on decentralized AI-based energy systems (Márquez and Gonzalo 2021). These systems vary from utility-scale, centralized models to a more decentralized setup where community-sized wind farms and decentralized microgrids operate independently of their generation, storage and distribution (Zhao et al. 2024). AI plays a key role in making the systems operate as optimally as possible by monitoring the local weather pattern, load profile and grid status in real-time (Liu et al. 2022). As these microgrids become more intelligent, they enhance energy access in off-grid locations, reduce transmission losses and provide greater resilience from power outages (Tsai et al. 2023).

6.7.1. *AI and blockchain synergy in wind markets*

Maybe the most thrilling convergence of technology is between AI and blockchain technology in enabling wind power transactions. Blockchain is an immutable, transparent ledger technology ideally placed to authenticate peer-to-peer energy transactions, and AI optimizes timing, price and demand forecasting (Kamisetty 2022). Together, these technologies would make self-governing energy markets possible, in which wind producers and consumers interact directly (Zhang et al. 2024). For example, AI can be used to calculate the optimal time to sell or hold energy based on changing market feedback, while blockchain enables authentic proof and tracing of transactions. The overall infrastructure is not only enhancing market efficiency but also future proof such as with tokenized energy credits and community energy cooperatives (Ukoba et al. 2024b).

6.7.2. *Role in carbon credit markets and smart grids*

AI will also be instrumental in carbon credit market growth and automation (Simankov et al. 2024). With countries and firms accelerating decarbonization efforts, concrete evidence of reducing emissions is required. AI technology can compute operating data from wind farms to demonstrate avoided emissions with a high level of accuracy and offer auditable values for the creation of carbon credits (Elkelawy et al. 2024). These verifications by AI can be implemented in real time within the world's carbon markets, enhancing the speed of value creation for wind power projects and enabling quicker monetization of environmental benefits (Li et al. 2025).

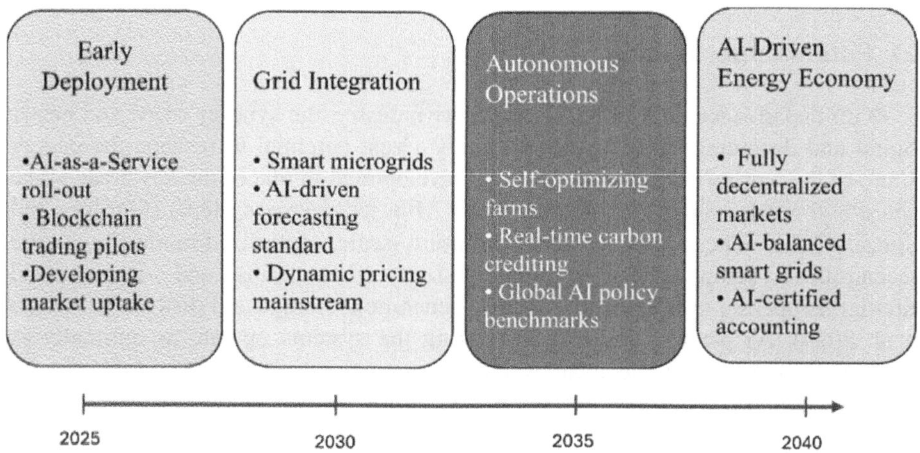

Figure 6.3. *Future roadmap for AI adoption in wind energy: 2025–2040 (phased projections). Sources: IRENA 2023 Innovation Outlook on AI & Big Data; Bloomberg NEF 2024 Digitalization & AI in Renewables Market Outlook; IEA 2024 Digital Demand-Driven Electricity Networks Roa\dmap; GWEC 2024 Global Wind Report; World Bank ESMAP 2023 Mini-Grids Market Outlook & AI Control Systems; IEEE P2418.5-2023 Dynaamic Pricing Guidelines; DNV 2024 Energy Transition Outlook – Digital Twins & Autonomous Turbines; Goldman Sachs 2024 Carbonomics Real-Time Carbon Accounting; EC-JRC 2023 AI for the Energy System 2040 Vision; IEA/IRENA 2023 Net-Zero Roadmap 1.5 °C Scenario – Digital & Market Enhancers. For a color version of this figure, see www.iste.co.uk/kumar/windpower2.zip*

In addition, AI deployment in smart grids will transform grid operation and demand response. With end-to-end real-time visibility of generation, use and storage of energy, AI can enable grid operators to manage frequency stability, load balance and schedule supply from multiple sources of renewables (Wei et al. 2021). Wind power, whose very nature is volatility, can particularly benefit from AI-driven coordination that continuously varies dispatch schedules, prevents overloads and coordinates backup systems in a perfectly coordinated way (Chen et al. 2021). In the longer term, there is a dream of an autonomous grid where AI maximizes the flow of energy with a minimal human touch.

In short, the future of AI in wind energy is much broader than existing operational improvement. As the industry adopts decentralization, market liberalization and environmental responsibility, AI will be both the driver and conductor of this change. Its fusion with technologies such as blockchain, its use in carbon trading and its central position in smart grid infrastructure make AI a core tool for the next generation of clean, resilient and democratized energy systems.

6.8. Conclusion

AI in the wind industry is not a wish of the future, but an economic and market-relevant reality of today. AI technologies have transformed the way wind farms generate and sell electricity through applications such as performance enhancement and predictive maintenance, sophisticated forecasting and automated trading. Economically, AI prevents running and capital costs through the implementation of data-informed decision-making, reducing downtime and optimizing the lifespan of core components. This translates to tangible reductions in the LCOE, consequently rendering wind power cheaper and viable as compared to energy from fossils fuels. In addition, AI enhances financial stability by accurate energy forecasting, allowing producers to perform a more effective role in changing markets, reduce risk and increase the ROI. All these combined redefine the investment rationale of wind projects, making them more appealing to institutional investors, governments and private developers alike. Besides financial benefits, AI can drive market forces to an all-time high in terms of sophisticated interaction with free-wheeling, decentralized energy markets. Electricity production from wind can now respond to price signals, maximize energy bidding strategy and even engage in peer-to-peer trading with block chain-supported platforms, all facilitated by AI-powered automation and analytics. With AI as a catalyst for sustainable growth, with the world moving rapidly towards decarbonization and energy autonomy, AI comes to be a driver towards the realization of these global goals. With enhanced operational reliability, minimized losses in energy and optimization of market efficiency, AI supports the drive of grand environmental sustainability and energy justice goals. Its use in carbon credit generation, smart grid operation and self-sustaining microgrids also evokes the vision for resilient and responsive power systems. Significantly, the ability of AI to transform variability in the wind into a manageable resource flip what was formerly a weakness into an asset. This reframing of the utility of wind power positions it as a cornerstone of next-generation clean energy systems.

Briefly, AI is not just an add-on technology within the wind power system but an inherent impetus of industry transformation. Its influence spans the entire width of wind power development, from engineering and economics to markets and policy. As digital technologies continue to become mature and more cost-effective, the application of AI will increasingly become indispensable for enlarging wind energy globally. For its full realization of the transformational potential of AI, however, fundamental investments, intersectoral collaboration and capacity development are needed. With proper policy and strategies in place, AI will be able to enable a new generation of smart, clean and economically feasible wind power, while enabling a low-carbon and efficient transition economy globally.

6.9. References

Acaroğlu, H. and Márquez, F.P.G. (2021). Comprehensive review on electricity market price and load forecasting based on wind energy. *Energies*, 14(22), 7473. doi: 10.3390/en14227473.

Al-Othman, A., Tawalbeh, M., Martis, R., Dhou, S., Orhan, M., Qasim, M., Olabi, A.G. (2021). Artificial intelligence and numerical models in hybrid renewable energy systems with fuel cells: Advances and prospects. *Energy Conversion and Management*, 253, 115154. doi: 10.1016/j.enconman.2021.115154.

Barbosa, N.B., Nunes, D.D.G., Santos, A.Á.B., Machado, B.A.S. (2023). Technological advances on fault diagnosis in wind turbines: A patent analysis. *Applied Sciences*, 13(3), 1721. doi: 10.3390/app13031721.

Behara, R.K. and Saha, A.K. (2022). Artificial intelligence control system applied in smart grid integrated doubly fed induction generator-based wind turbine: A review. *Energies*, 15(17), 6488. doi: 10.3390/en15176488.

BloombergNEF (2024). Digitalisation & AI in Renewables 2024 Market Outlook. London.

Bošnjaković, M., Martinović, M., Đokić, K. (2025). Application of artificial intelligence in wind power systems. *Applied Sciences*, 15(5), 2443. doi: 10.3390/app15052443

Chang, V., Chen, Y., Zhang, Z., Xu, Q.A., Baudier, P., Liu, B.S. (2021). The market challenge of wind turbine industry-renewable energy in PR China and Germany. *Technological Forecasting and Social Change*, 166, 120631. doi: 10.1016/j.techfore.2021.120631.

Chatterjee, J. and Dethlefs, N. (2021b). Scientometric review of artificial intelligence for operations & maintenance of wind turbines: The past, present and future. *Renewable and Sustainable Energy Reviews*, 144, 111051. doi: 10.1016/j.rser.2021.111051.

Chen, C., Hu, Y., Karuppiah, M., Kumar, P.M. (2021a). Artificial intelligence on economic evaluation of energy efficiency and renewable energy technologies. *Sustainable Energy Technologies and Assessments*, 47, 101358. doi: 10.1016/j.seta.2021.101358.

Chen, X., Eder, M.A., Shihavuddin, A., Zheng, D. (2021b). A human-cyber-physical system toward intelligent wind turbine operation and maintenance. *Sustainability*, 13(2), 561. doi: 10.3390/su13020561.

Dehshiri, S.S.H. and Firoozabadi, B. (2024). A multidisciplinary approach to select wind turbines for power-hydrogen production: Energy, exergy, economic, environmental under uncertainty prediction by artificial intelligence. *Energy Conversion and Management*, 310, 118489. doi: 10.1016/j.enconman.2024.118489.

DNV (2024). Energy Transition Outlook 2024: Digital Twins & Autonomous Wind Turbines. Oslo.

Elkelawy, M., Atta, Z.A., Seleem, H. (2024). Technological advances, efficiency optimization, and challenges in wind power plants: A comprehensive review. *Pharos Engineering Science Journal*, 1(1), 57–65. doi: 10.21608/pesj.2025.344906.1006.

Elmousalami, H., Alnaser, A.A., Hui, F.K.P. (2025b). Sustainable AI-driven wind energy forecasting: Advancing zero-carbon cities and environmental computation. *Artificial Intelligence Review*, 58(6). doi: 10.1007/s10462-025-11191-0.

Energy Information Administration (2021). AI applications in energy forecasting [Online]. Available at: https://www.eia.gov.

European Commission Joint Research Centre (2023). AI for the energy system 2040 vision. Luxembourg.

Franki, V., Majnarić, D., Višković, A. (2023). A comprehensive review of artificial intelligence (AI) companies in the power sector. *Energies*, 16(3), 1077. doi: 10.3390/en16031077.

GE Renewable Energy (n.d.). Artificial intelligence and digital solutions [Online]. Available at: https://www.ge.com/renewableenergy.

Global Wind Energy Council (2024). Global Wind Report 2024. Brussels.

Goldman Sachs (2024). Carbonomics: Real-time carbon accounting platforms. Research Report. New York.

Harrison-Atlas, D., Glaws, A., King, R.N., Lantz, E. (2024). Artificial intelligence-aided wind plant optimization for nationwide evaluation of land use and economic benefits of wake steering. *Nature Energy*, 9(6), 735–749. doi: 10.1038/s41560-024-01516-8.

IEEE Power and Energy Society (2023). Dynamic pricing for variable renewables – 2030 guidelines. IEEE Std P2418.5.

International Energy Agency (n.d.). Artificial intelligence and energy systems [Online]. Available at: https://www.iea.org.

International Energy Agency (2021a). Technology roadmap: Artificial intelligence in energy systems [Online]. Available at: https://www.iea.org/reports/artificial-intelligence.

International Energy Agency (2021b). Digitalization and energy [Online]. Available at: https://www.iea.org/reports/digitalisation-and-energy.

International Energy Agency (2024). Digital Demand-Driven Electricity Networks (Grid-Edge) Roadmap 2024. Paris.

International Energy Agency and International Renewable Energy Agency (2023). Net-Zero Roadmap – 1.5°C Scenario: Digital & Market Enhancers. Paris/Abu Dhabi.

International Renewable Energy Agency (n.d.). Publications [Online]. Available at: https://www.irena.org/publications.

International Renewable Energy Agency (2020). Innovation outlook: Artificial intelligence and big data [Online]. Available at: https://www.irena.org/publications.

International Renewable Energy Agency (2023). Innovation outlook: Artificial intelligence and big data – 2023 update. Abu Dhabi.

Jiang, Z. and Liu, Z. (2021). Can wind power policies effectively improve the productive efficiency of Chinese wind power industry? *International Journal of Green Energy*, 18(13), 1339–1351. doi: 10.1080/15435075.2021.1904407.

Kamisetty, A. (2022). AI-driven robotics in solar and wind energy maintenance: A path toward sustainability. *Asia Pacific Journal of Energy and Environment*, 9(2), 119–128. doi: 10.18034/apjee.v9i2.784.

Khalid, M. (2024). Energy 4.0: AI-enabled digital transformation for sustainable power networks. *Computers & Industrial Engineering*, 193, 110253. doi: 10.1016/j.cie.2024.110253.

Khan, K.I. and Nasir, A. (2023). Application of artificial intelligence in solar and wind energy resources: A strategy to deal with environmental pollution. *Environmental Science and Pollution Research*, 30(24), 64845–64859. doi: 10.1007/s11356-023-27038-6.

Lazard (2023). Levelized cost of energy and storage – version 17.0 [Online]. Available at: https://www.lazard.com/perspective/lcoe2023.

Li, B. (2023). The role of financial markets in the energy transition: An analysis of investment trends and opportunities in renewable energy and clean technology. *Environmental Science and Pollution Research*, 30(43), 97948–97964. doi: 10.1007/s11356-023-29014-6.

Li, L., Wen, J., Li, Y., Mu, Z. (2025). Supply chain challenges and energy insecurity: The role of AI in facilitating renewable energy transition. *Energy Economics*, 108378. doi: 10.1016/j.eneco.2025.108378.

Lipu, M.S.H., Miah, M.S., Hannan, M.A., Hussain, A., Sarker, M.R., Ayob, A., Saad, M.H.M., Mahmud, M.S. (2021). Artificial intelligence based hybrid forecasting approaches for wind power generation: Progress, challenges and prospects. *IEEE Access*, 9, 102460–102489. doi: 10.1109/access.2021.3097102.

Liu, Z., Sun, Y., Xing, C., Liu, J., He, Y., Zhou, Y., Zhang, G. (2022). Artificial intelligence powered large-scale renewable integrations in multi-energy systems for carbon neutrality transition: Challenges and future perspectives. *Energy and AI*, 10, 100195. doi: 10.1016/j.egyai.2022.100195.

Loi, N.T., Huyen, N.T.T., Thanh, T.T., Ha, L.T. (2023). Network interlinkages between artificial intelligence and green energy dynamics during the War in a Pandemic: An application of Bayesian vector heterogeneous autoregressions. *Environmental Challenges*, 13, 100796. doi: 10.1016/j.envc.2023.100796.

Machado, J.T.M. and De Andrés, M. (2022). Implications of offshore wind energy developments in coastal and maritime tourism and recreation areas: An analytical overview. *Environmental Impact Assessment Review*, 99, 106999. doi: 10.1016/j.eiar.2022.106999.

Márquez, F.P.G. and Gonzalo, A.P. (2021). A comprehensive review of artificial intelligence and wind energy. *Archives of Computational Methods in Engineering*, 29(5), 2935–2958. doi: 10.1007/s11831-021-09678-4.

Mauludin, M.S., Khairudin, M., Asnawi, R., Mustafa, W.A., Toha, S.F. (2024). The advancement of artificial intelligence's application in hybrid solar and wind power plant optimization: A study of the literature. *Journal of Advanced Research in Applied Sciences and Engineering Technology*, 50(2), 279–293. doi: 10.37934/araset.50.2.279293.

Mhlanga, D. (2023). Artificial intelligence and machine learning for energy consumption and production in emerging markets: A review. *Energies*, 16(2), 745. doi: 10.3390/en16020745.

Mitchell, D., Blanche, J., Harper, S., Lim, T., Gupta, R., Zaki, O., Tang, W., Robu, V., Watson, S., Flynn, D. (2022). A review: Challenges and opportunities for artificial intelligence and robotics in the offshore wind sector. *Energy and AI*, 8, 100146. doi: 10.1016/j.egyai.2022.100146.

Msigwa, G., Ighalo, J.O., Yap, P. (2022). Considerations on environmental, economic, and energy impacts of wind energy generation: Projections towards sustainability initiatives. *The Science of the Total Environment*, 849, 157755. doi: 10.1016/j.scitotenv.2022.157755.

Nazir, M.S., Wang, Y., Bilal, M., Abdalla, A.N. (2022). Wind energy, its application, challenges, and potential environmental impact. In *Handbook of Climate Change Mitigation and Adaptation*, Lackner, M., Sajjadi, B., Chen, W.-Y. (eds). Springer, Cham. doi: 10.1007/978-3-030-72579-2_108.

Ohalete, N.N.C., Aderibigbe, N.A.O., Ani, N.E.C., Ohenhen, N.P.E., Daraojimba, N.D.O., Odulaja, N.B.A. (2023). AI-driven solutions in renewable energy: A review of data science applications in solar and wind energy optimization. *World Journal of Advanced Research and Reviews*, 20(3), 401–417. doi: 10.30574/wjarr.2023.20.3.2433.

Onwusinkwue, N.S., Osasona, N.F., Ahmad, N.I.A.I., Anyanwu, N.A.C., Dawodu, N.S.O., Obi, N.O.C., Hamdan, N.A. (2024). Artificial intelligence (AI) in renewable energy: A review of predictive maintenance and energy optimization. *World Journal of Advanced Research and Reviews*, 21(1), 2487–2799. doi: 10.30574/wjarr.2024.21.1.0347.

Rane, N., Choudhary, S., Rane, J. (2023). Leading-edge artificial intelligence (AI)-powered financial forecasting for shaping the future of investment strategies. *SSRN Electronic Journal*. doi: 10.2139/ssrn.4640828.

Rhatrif, A.E., Bouihi, B., Mestari, M. (2024). Challenges and limitations of artificial intelligence implementation in modern power grid. *Procedia Computer Science*, 236, 83–92. doi: 10.1016/j.procs.2024.05.008.

Rusilowati, U., Ngemba, H.R., Anugrah, R.W., Fitriani, A., Astuti, E.D. (2024). Leveraging AI for superior efficiency in energy use and development of renewable resources such as solar energy, wind, and bioenergy. *International Transactions on Artificial Intelligence (ITALIC)*, 2(2), 114–120. doi: 10.33050/italic.v2i2.537.

Shahid, A., Plaum, F., Korõtko, T., Rosin, A. (2024). AI technologies and their applications in small-scale electric power systems. *IEEE Access*, 12, 109984–110001. doi: 10.1109/access.2024.3440067.

Siemens Gamesa Renewable Energy (n.d.). Digitalization and artificial intelligence in wind power [Online]. Available at: https://www.siemensgamesa.com.

Simankov, V., Buchatskiy, P., Kazak, A., Teploukhov, S., Onishchenko, S., Kuzmin, K., Chetyrbok, P. (2024). A solar and wind energy evaluation methodology using artificial intelligence technologies. *Energies*, 17(2), 416. doi: 10.3390/en17020416.

Sofian, A.D.A.B.A., Lim, H.R., Munawaroh, H.S.H., Ma, Z., Chew, K.W., Show, P.L. (2024). Machine learning and the renewable energy revolution: Exploring solar and wind energy solutions for a sustainable future including innovations in energy storage. *Sustainable Development*, 32(4), 3953–3978. doi: 10.1002/sd.2885.

Stecyk, A. and Miciuła, I. (2023). Harnessing the power of artificial intelligence for collaborative energy optimization platforms. *Energies*, 16(13), 5210. doi: 10.3390/en16135210.

Sun, Y. and Han, W. (2025). A review of enhancing wind power with AI: Applications, economic implications, and green innovations. *Digital Economy and Sustainable Development*, 3(1). doi: 10.1007/s44265-025-00059-4.

Thapa, N. (2022). AI-driven approaches for optimizing the energy efficiency of integrated energy system. Master's Thesis, University of Vaasa.

Tsai, W., Hong, C., Tu, C., Lin, W., Chen, C. (2023). A review of modern wind power generation forecasting technologies. *Sustainability*, 15(14), 10757. doi: 10.3390/su151410757.

Ukoba, K., Olatunji, K.O., Adeoye, E., Jen, T., Madyira, D.M. (2024b). Optimizing renewable energy systems through artificial intelligence: Review and future prospects. *Energy & Environment*, 35(7), 3833–3879. doi: 10.1177/0958305x241256293.

U.S. Department of Energy (n.d.). How wind turbines work [Online]. Available at: https://www.energy.gov/eere/wind/how-wind-turbines-work.

Vestas (n.d.). AI and digital solutions for wind energy [Online]. Available at: https://www.vestas.com.

Wang, J., Cheng, L., Feng, L., Lin, K., Zhang, L., Zhao, W. (2023). Tracking and predicting technological knowledge interactions between artificial intelligence and wind power: Multimethod patent analysis. *Advanced Engineering Informatics*, 58, 102177. doi: 10.1016/j.aei.2023.102177.

Wei, Y., Ye, Q., Ding, Y., Ai, B., Tan, Q., Song, W. (2021). Optimization model of a thermal-solar-wind power planning considering economic and social benefits. *Energy*, 222, 119752. doi: 10.1016/j.energy.2021.119752.

WindEurope (2019). Wind energy in Europe in 2019: Trends and statistics [Online]. Available at: https://windeurope.org.

World Bank ESMAP (2023). Mini-grids for half a billion people – Market outlook & AI control systems. Washington, D.C.

Yousef, L.A., Yousef, H., Rocha-Meneses, L. (2023). Artificial intelligence for management of variable renewable energy systems: A review of current status and future directions. *Energies*, 16(24), 8057. doi: 10.3390/en16248057.

Yu, M., Niu, D., Gao, T., Wang, K., Sun, L., Li, M., Xu, X. (2023). A novel framework for ultra-short-term interval wind power prediction based on RF-WOA-VMD and BiGRU optimized by the attention mechanism. *Energy*, 269, 126738. doi: 10.1016/j.energy.2023.126738.

Zhang, X., Khan, K., Shao, X., Oprean-Stan, C., Zhang, Q. (2024). The rising role of artificial intelligence in renewable energy development in China. *Energy Economics*, 132, 107489. doi: 10.1016/j.eneco.2024.107489.

Zhao, L., Nazir, M.S., Nazir, H.M.J., Abdalla, A.N. (2022). A review on proliferation of artificial intelligence in wind energy forecasting and instrumentation management. *Environmental Science and Pollution Research*, 29(29), 43690–43709. doi: 10.1007/s11356-022-19902-8.

Zhao, C., Dong, K., Wang, K., Nepal, R. (2024). How does artificial intelligence promote renewable energy development? The role of climate finance. *Energy Economics*, 133, 107493. doi: 10.1016/j.eneco.2024.107493.

7

AI for Policy and Regulatory Frameworks in Wind Power

The goal of switching to sustainable sources of energy, especially wind energy, is seen globally as one to combat climate change and limit dependency on fossil fuels. In spite of this, deploying wind energy systems is intricately shaped by complex policy landscapes and regulatory frameworks. In this chapter, we discuss how artificial intelligence (AI) can influence policymaking and regulatory evolutions in the wind power domain by improving, automating and optimizing the instruments of policy and regulatory systems. AI-based systems can also facilitate data-driven policymaking, risk-sensitive regulation, the forecasting of environmental impacts and real-time monitoring of wind farm functioning. Through machine learning, natural language processing and data analytics, AI has the potential to optimize bureaucratic workflows, predict regulatory risks and enable flexible governance models. This chapter provides an overview of current challenges in wind energy regulatory mechanisms, followed by case studies where AI has been implemented into renewable policy planning, before concluding with the potential future opportunities of AI in regulatory ecosystems. This approach allows a more dynamic, transparent and effective implementation of sustainable wind energy through the integration and coupling of AI in both policy design and regulatory framework delivery.

7.1. Introduction

Wind power is leading the charge towards a low-carbon future as the world continues to ramp up the transition to renewable energy. However, the rapid growth of wind electricity infrastructure poses major challenges for policymakers and regulators. Whether it is balancing grid stability or environmental and socio-economic impacts, regulatory frameworks need to keep pace with technology development.

Chapter written by Suman CHAHAR and Komal MISHRA.

In this context, new artificial intelligence (AI) tools are emerging with new opportunities to better support policy design, regulatory compliance or decision-making in the wind power sector. Machine learning (ML), natural language processing (NLP) and predictive analytics are AI technologies that analyze vast datasets in real time. This helps regulators to adopt more adaptive and evidence-based approaches. In fact, AI can help predict wind patterns, energy yields, environmental risks and even regulatory bottlenecks. In addition, AI-enabled simulations and scenario analysis tools allow policymakers to explore the implications of diverse regulatory approaches in complex and uncertain environments. This chapter examines the intersection of AI and regulatory policy in the development of wind power. It explores the role of AI in fostering agile, transparent and efficient regulatory frameworks as well as the need to respond to rapidly changing technology, environmental and market forces. It also examines the ethical, legal and governance issues that need to be resolved when AI is deployed to support policymaking processes. The chapter studies stakeholders (including governments, regulatory agencies, energy developers and researchers) seeking to leverage AI for the responsible acceleration of wind power deployment. All things considered, it highlights the possibilities of AI, not simply as a technological facilitator, but as a strategic ally in the development of a more intelligent, more sustainable energy policy for the road ahead.

7.2. Challenges in current policy and regulatory frameworks

7.2.1. *Completely static and rigid regulations*

The current policy landscape for wind power is fraught with challenges, one of which is the inherent inflexibility of the existing regulatory framework. Previous policies were written based on clean energy paradigms that are poorly suited to the fluid nature of renewable energy technologies. Wind power generation, for instance, is inherently variable and site-specific, but existing regulations tend to take a one-size-fits-all approach that fails to consider local wind patterns, grid capacities or environmentally sensitive areas. Such inflexible rules can slow project approvals, stifle innovation and deter private investment. Additionally, lengthy policy revision cycles cause regulations to fall behind technological developments. The pace of deployment remains constrained by rigid grid codes, antiquated permitting processes, and persistent bureaucratic hurdles. As such, there is an urgent demand for information-driven, agile and dynamic regulatory mechanisms that can keep pace with the constantly evolving wind energy ecosystem. AI can solve this challenge through adaptive regulations based on predictive modeling and real-time analytics.

But while such systems are not yet widely embedded, the inertia will continue to bite on the scalability and sustainability of wind power development. Another key challenge is the limited use of data and advanced analytics in the policymaking process. Current regulatory practices depend heavily on static datasets, human experience and past trends with an outdated understanding of wind energy system dynamics. The resulting decisions about site approvals, subsidy allotments, environmental impact assessments (EIA) and even grid integration approaches are therefore suboptimal. However, the lack of real-time data integration means policymakers are not able to react quickly to shifts in wind patterns, energy demands or environmental constraints. In addition, siloed data systems across diverse governmental and private institutions obstruct detailed analysis and coordinated action. AI can be a key figure in overcoming those limitations through real-time data processing, forecasting and scenario simulations. Yet, the underlying problem remains: existing frameworks do not possess the digital infrastructure and policy mandates to implement data-centric approaches. Without this shift, regulatory bodies risk lagging behind the pace of technological development, slowing down the global transition to renewable energy (Chahar et al. 2024).

7.2.2. *Poor integration with grid management systems*

Persistent flaws in wind power policy create regulatory inconsistencies that undermine the stability of grid management systems. As the most variable renewable energy source, wind energy can be weather-dependent and must be constantly facilitated through dynamic integration into the electrical grid in order to keep the energy system stable and efficient. But many existing frameworks are rooted in legacy grid models that were designed around centralized, fossil-fuel-based generation. This mismatch can cause issues such as wind power curtailment, frequency instability and inefficient energy dispatch. Regulatory bottlenecks also slow investments in smart grids and energy storage, which play a central role in balancing intermittent supply. In addition, there are often little to no regulatory requirements or incentives for utilities to implement AI-infused grid forecasting tools and adaptive load balancing strategies. Without such integration, the operators of these systems struggle to plan for the grid capacity, forecast power generation or optimize demand-side responses effectively. Consequently, even if wind farms can function up to capacity, they cannot do so because they do not consider grid infrastructure. Bridging this gap calls for strategic policies that support AI-powered grid intelligence, real-time monitoring and predictive load balancing; these are concepts that have not yet entered the playbook for most existing frameworks (Bošnjaković et al. 2025).

7.2.3. *Permission latency and environmental approvals*

Wind energy projects often experience year-long delays in regard to permitting and EIA procedures. Although it is necessary to have environmental safeguards, the regulatory procedures surrounding them are frequently outdated, opaque and inconsistent throughout jurisdictions. Not only do these delays extend project costs they also deter investors and developers. Permitting procedures can take years, even in wind-rich areas, because of manual paperwork, responsibilities split across different agencies and a lack of digital tools to speed the reviews along. Moreover, EIAs often use static models, with real-time data or predictive tools not being harnessed to more accurately gauge wildlife impact, noise levels or land-use concerns. This helps increase community opposition and legal challenges. AI offers a path to streamlining these processes through sophisticated environmental modeling, automated document review, and multi-scenario simulations. However, until regulators formally adopt these tools, bureaucratic hurdles will continue to stall progress. Implementing policy reforms to standardize and digitalize permitting procedures would reduce administrative bottlenecks and provide clear AI-supported guidelines for the implementation of EIAs. The absence of such reforms continues to inhibit timely deployment of wind power (Lüthi and Prässler 2011).

7.2.4. *Lack of adequate stakeholder engagement and transparency*

At the local level, this results in mistrust, opposition and conflict, whereas current regulatory frameworks are often not sufficient to ensure meaningful stakeholder participation. The relevance of the affected community in wind farm developments, be it through visual impact, land use or generalized environmental concerns, is often only considered too late or, at best, in a limited capacity. This reactive strategy leads to resistance and can cause expensive hold-ups or project abandonment. In addition, interest-oriented policies and decisions are often made in closed societies which take place behind closed doors, devoid of participatory decision-making processes or open communication. There are also no accessible platforms where the public, researchers and interest groups can access and discuss data, models and impact assessments. AI and digital platforms could enhance transparency by visualizing data and forecasting impacts, as well as facilitating real-time feedback from communities. However, such tools are infrequently called for or even supported in existing frameworks.

In regard to policy changes for improving regulatory transparency and engagement, it will take more than technological tools to achieve this. It will also require policy shifts towards more inclusive governance models. Until that happens, weak stakeholder involvement will undermine the democratic and equitable development of the wind energy sector (Danish and Senjyu 2023a).

7.2.5. *Mismatched incentive structures*

Several existing policy measures, such as subsidies, feed-in tariffs and tax-exemptions are not sufficiently aligned with the long-term targets of wind energy expansion or grid optimization. Such incentives tend to be short term and too broad, or do not correlate with actual performance on a project, leading to market inefficiencies and missed opportunities. For instance, a flat subsidy rate may lead to the overdevelopment of low-wind areas or may not compensate for innovations in storage, efficiency or AI-based forecasting solutions. In addition, capacity rather than resilience may be a key factor in regulatory schemes, failing to fully capture the value of systems that can make electric grids smarter and thus increase their ability to respond to deployment challenges and operate harmoniously. The absence of performance-linked incentives also stymies the adoption of newer technologies, including AI-based predictive maintenance or adaptive dispatch of energy systems. AI can provide more nuanced information on project viability and performance to allow for flexible and responsive incentive structures. However, the policy landscape we have in place today lacks the analytical infrastructure and adaptive capacity to realize such models. The repowering of these frameworks with data-driven, performance-based incentives is critical to unleashing the full wind potential in a cost-effective manner and help future-proof the energy supply on a physical as well as commercial level (Danish and Senjyu 2023b)

7.2.6. *Fragmentation of international and subnational policies*

The wind energy sector faces similar hurdles, with policy harmonization failing to extend between borders and between different levels of government. Differences in regulatory standards, reporting requirements and approval processes can lead to confusion, compliance costs and investment risks, particularly for multinational developers and cross-border infrastructure projects. One example is having completely different grid codes, environmental requirements and data sharing obligations in two neighboring countries, or two provinces of the same country. This regulatory fragmentation hampers innovation, reduces economies of scale and weakens the promise of shared AI-based solutions for planning, forecasting and risk assessment. The absence of common digital platforms and interoperable regulatory systems also forestalls collaborative modeling or data pooling, which are critical for training successful AI systems. Many frameworks still operate under silos, while some regions are making strides towards alignment (e.g. through regional energy markets or intergovernmental agreements). A more global scalability of wind energy requires these facets to work in tandem in terms of interoperability, shared standards and collaboration between AI systems.

7.3. Overview of AI technologies relevant to policy and regulation

AI includes an extensive set of technologies that can usefully augment the design, operation and reviewing of an effective policy and regulatory framework in wind energy. Such technologies allow for governance models that are more responsive, data-driven and adaptable, and hence more capable of handling the complexity and dynamism of the renewable energy sector. Below are some key AI technologies that are very relevant.

7.3.1. *Machine learning (ML)*

ML is a subset of AI that focuses on building algorithms that can learn from and make predictions on data. ML is particularly important in predictive modeling and policy optimization in wind energy regulation processes. For example, ML can be trained on past wind speeds, the times requests were made on the grid and turbine efficiencies, and predict future energy outputs with high accuracy. This predictive ability will enable policymakers to create flexible regulatory tools that can be adjusted according to fluctuations in energy supply. ML is also leveraged for identifying bottlenecks in approval processes, such as analyzing patterns in permitting timelines and stakeholder responsiveness. ML models and techniques that learn from labeled data (supervised ML models) can help evaluate the success of previous regulatory interventions, while methods based on unsupervised learning can be used to cluster regions or projects by compliance risk factors or performance metrics. Reinforcement learning, an advanced subbranch of ML, can be used to push the evolutionary path by simulating the regulatory strategies in timely environments and improving upon the regulatory design. Ultimately, ML provides regulators with data-driven insights gained from large and complex datasets allowing for smarter and quicker decision-making in wind energy regulation (Chahar et al. 2025).

7.3.2. *Natural language processing (NLP)*

Natural language processing (NLP) is a branch of AI that is concerned with bridging the gap between human communication and machine understanding. This has broad applications for analyzing policy documents, public consultations, legal texts and regulatory submissions in the area of wind power. Policymaking is frequently oriented around studying thousands of pages of technical reports, environmental impact statements and public comment processes that must be undertaken, which are lengthy and prone to human error. NLP tools can help automate the summarization of the top three or five key insights or areas that are most important to the organization, highlight inconsistencies and track the same issues that are repeatedly raised in very broad terms in stakeholder feedback. Sentiment analysis, a sub-field of NLP, can assess public sentiment towards proposed wind projects, helping regulators harmonize their pursuit of development with social acceptance. NLP can also help in harmonizing language in regulations across jurisdictions by identifying semantic discrepancies or redundancies in legislation. In multilingual or international contexts, translation and summation tools powered by NLP promote wider accessibility and transparency of content.

Furthermore, the key to real progress is embedding NLP capabilities into regulatory decision support platforms that can scan thousands of policy drafts, legal precedents and public discourse in real time, greatly alleviating the administrative burden while developing inclusiveness and continuity in regulatory processes.

7.3.3. *Predictive analytics*

Predictive analytics is the process of using statistical methods, AI and historical data to make informed predictions about the likelihood of future trends, patterns or outcomes. Scenario modeling, risk assessment and regulatory planning are areas where predictive analytics holds a particular value in wind power policy and regulation. For instance, regulators might use predictive models to inform them of looming grid congestion, project the long-term costs of subsidy schemes or even model the environmental consequence of new wind deployments. These models integrate data from weather forecasts, energy markets, social trends and environmental indicators to deliver integrated, forward-looking insights. Similarly, predictive analytics also discerns possible high-risk zones that could be vulnerable to non-compliance or community resentment, which allows for preventive policy alteration. By embedding these forecasts into the regulatory cycle, decision-makers can better simulate a range of policy impacts under alternative assumptions, for instance, changes in carbon pricing, shifts in energy demand or climate events, and achieve stronger, more resilient regulation. Coupled with AI techniques such as ML and NLP, predictive analytics can be further enhanced, thus making it a foundational alchemy for data-driven and adaptive policy development in wind energy.

7.3.4. *Decision support systems: DSS*

Decision support systems powered by AI are intelligent platforms created to help policymakers assess complex trade-offs and make better decisions. They synthesize real-time data with simulations, stakeholder inputs and optimization models to deliver actionable recommendations for decision-makers. In the wind power regulation arena, DSS can help to zone decisions by evaluating terrain data, noise pollution models, biodiversity impact and energy yield at the same time. They also help align policy goals with technical constraints by modeling a range of regulatory options from feed-in tariffs, tax incentives to carbon caps and evaluating their expected effects on energy generation, emissions and cost-effectiveness. The DSS platform often has interactive dashboards, which enable regulators to view "what if" scenarios and visualize the outcomes of policy implementation. They can, for example, evaluate what accelerating the threshold for community consultations would mean for delaying a project, or how subsidizing more high-yield zones could influence the energy security of the high-yield zones being created. With the integration of AI capabilities, modern DSS can improve over time by learning from outcomes to provide more refined recommendations and adapt to new incoming data. This means they can be an important instrument in adopting transparency and improving the speed of regulatory processes in the rapidly changing reality of wind energy.

7.3.5. *Expert systems and rule-based AI*

Expert systems are rule-based AI applications that act as human experts by simulating decision-making capabilities. Such systems are especially useful in policy contexts characterized by complex criteria, legal constraints and conditional logic. An example of how an expert system could be useful for the wind-power-regulated sector is it can systematically automate the processing of permit applications by checking requirements. It can also check whether they are being followed according to technical standards and environmental requirements. Expert systems could, for example, assess whether a proposed wind farm complies with rules on turbine spacing, wildlife protection zones or noise limits, and they could give feedback in a uniform and impartial manner. These systems rely on extensive knowledge bases, up-to-date regulatory codes, legal precedents and scientific models and are responsible for providing high-quality advice or decisions. There are AI use cases powered by rule-based models, for example, automatic drafting of regulatory documents by assembling clauses either through user inputs or defined policy templates. When integrated into regulatory platforms, expert systems can provide real-time decision support to both regulators and developers, which can mitigate administrative hold-ups and human error. They do not learn such as ML systems, but their transparency and predictability mean they are ideal in formal policy environments where traceability and legal compliance are essential.

7.3.6. *Knowledge graphs and semantic AI*

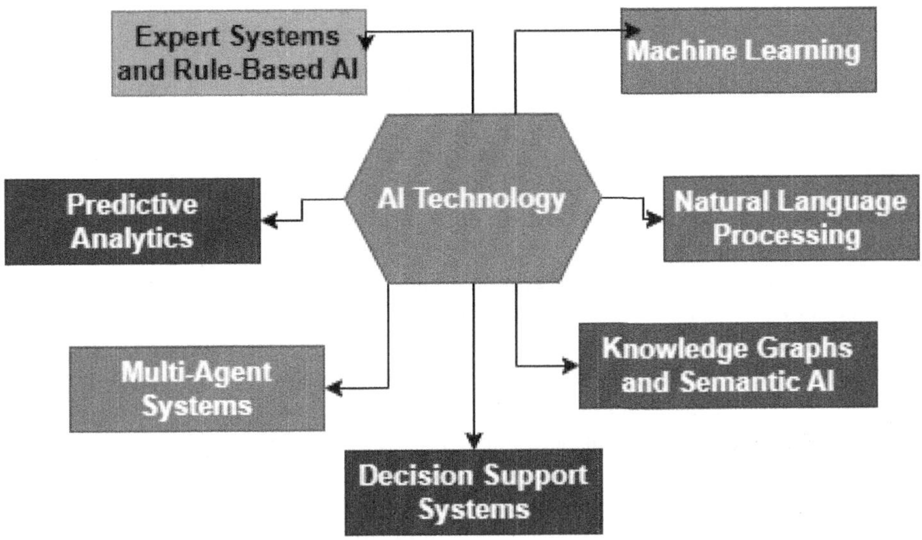

Figure 7.1. *Overview of AI technologies relevant to policy and regulation. For a color version of this figure, see www.iste.co.uk/kumar/windpower2.zip*

Knowledge graphs are a type of AI that model complex interrelations of entities such as policies, stakeholders, legal requirements and infrastructure as data structures based on graphs. Knowledge graphs can facilitate the integration of fragmented wind power regulatory information from different departments, jurisdictions and datasets in order to improve wind power regulation. For example, a knowledge graph may associate a wind project's location with relevant zoning laws, protections for endangered wildlife in the area, historical community feedback and applicable environmental data. This semantic interlinking helps policymakers to quickly find relevant precedents, highlight conflicts between regulations and identify hidden dependencies in regulatory texts. Another important aspect of official recognition is the facilitation or conditioning of (inter)national harmonization in legal regulations, where semantic AI can highlight conceptual neighborhoods and gaps between two or more legal systems or regional legislative frameworks. This is why knowledge graphs improve transparency, enable real-time policy audits and empower dynamic decision-making when built into AI platforms. These are especially effective in multi-stakeholder environments where understanding a wide range of connected variables is needed. Still relatively untapped in energy regulation today, knowledge graphs have tremendous potential to reduce the friction associated with governance by making complex regulatory ecosystems more searchable, navigable and analyzable.

7.3.7. *AI in remote sensing and computer vision*

Computer vision, a type of AI that enables machines to understand visual information, is an emerging method used in conjunction with remote sensing technologies (e.g. drones, satellites), in order to improve regulatory scrutiny in the wind sector. Machines can use these tools to automatically monitor construction at wind farms and compliance with zoning limits, as well as environmental impacts such as deforestation or migratory bird patterns. AI models trained on image and video data are effective in detecting anomalies such as unauthorized land use, erosion or turbine damage, and triggering automated alerts for regulatory attention. For environmental agencies focusing on the health of their core ecosystems, AI-enhanced imagery offers an additional level of accuracy and recency of visual evidence that can be incorporated into the compliance reports and audits they perform. In policy planning, computer vision assists spatial analysis in selecting sites by analyzing land attributes, proximity to infrastructure and ecological sensitivity. More generally, visual data from remote sensing can vitalize environmental impact models, providing more fine-grain, real-time inputs than traditional surveys. "The combination of computer vision with regulatory frameworks not only enhances the accuracy and timeliness of enforcement but also lowers the cost and effort of manual inspections resulting in more scalable and efficient regulatory models" (Chahar et al. 2025).

7.3.8. *Multi-agent systems (MAS)*

MAS combine autonomous behavior with multiple agents that can gather information, model real-world phenomena and problem-solve as a group in a specified environment. Typical applications of MAS include modeling the interaction between different entities of interest such as energy producers, grid operators, regulators, local governments and communities in the context of wind power regulation. Agents are given rules, objectives and capabilities, so that real-world dynamics can be simulated under various policy scenarios. MAS, for instance, could be applied to assess how new subsidy schemes could affect investor behavior, how more stringent environmental regulations would impact project viability and what communities may do in response to various consultation approaches.

In this way, by observing how these agents' behaviors emerge, policymakers could identify any unintended or systemic bottlenecks they might not get to see in traditional modeling approaches more successfully. MAS also provide a safer environment to test adaptive regulatory strategies before implementation in the real world, thereby reducing risk and enhancing the quality of the decision. In decentralized, multi-actor contexts such as energy markets, interactions are dynamic and interdependent, making this technology particularly useful. As the governance of this rapidly changing technology becomes more complex, this chapter demonstrates that MAS represents a powerful framework to build regulatory tools which are resilient, inclusive and adaptive across scales.

7.4. Applications of AI in wind power policy and regulation

7.4.1. *Policy simulation and forecasting*

The minimum-distance technique used for value-added and risk measurement (e.g. within this chapter) hinges on ML data that builds state-of-the-art simulation models informed of long-term predictive capabilities in terms of the scenario horizon of various policy interventions within the domain of wind power. With historical data in hand, ML algorithms, for example, can analyze how different policy instrument subsidies, tax incentives, altitude projects and carbon pricing have affected wind power deployment, grid stability and investor behavior. These simulations enable policymakers to model future impacts under different economic and environmental scenarios. For example, predictive analytics can be used to model the impact of a change in feed-in tariffs on regional energy supply, land use or emission targets. AI-enabled simulations take into consideration nonlinear interactions and uncertainties, thus making them more resilient than conventional forecasting tools. These models enable risk-informed decision-making and the alignment of energy policies with broader national energy and climate goals. AI improves policy development, in return for reducing exploration time by informing which policy levers yield higher impact outcomes that lead to the strategic design of regulatory frameworks (Danish and Senjyu 2023a).

7.4.2. *Automation of permitting and compliance monitoring*

The permitting process for wind energy remains a significant bottleneck, slowed by layers of administrative friction and inconsistent regulatory oversight. AI has the potential to ease and automate many steps in this process. NLP systems can scan applications and pull out critical compliance flags, while rule-based expert systems can determine if a proposed project complies with zoning ordinances, environmental laws and health safety codes. Moreover, AI-enabled document classification and anomaly detection can be used to instantly flag any discrepancies or incomplete submissions. The operation of the already installed wind containers can be supported by Freisinn's AI tools and technologies such as remote sensing and computer vision to monitor and ensure compliance with required environmental parameters (e.g. proximity to protected areas, noise levels) and generate automatic alerts in case of inconsistencies. This lightens the human regulators' load, while enhancing precision and responsiveness in oversight. Thus, permitting is quicker, more transparent and more accurate – creating an easier path for responsible wind energy development.

7.4.3. *Adaptive movement of regulatory states*

Traditional regulatory regimes are inherently slow-moving and cumbersome to change. AI provides a way to dynamically improve data-driven adaptive regulation. For instance, real-time grid data, wind forecasts and market signals can be amassed and analyzed continuously by ML models to recommend timely regulatory updates. These include possible changes in production limits, curtailment protocols or balancing requirements as conditions evolve. Feedback loops powered by AI mean policies can react quickly to changes in energy supply, energy demand and environmental limitations. This is especially true for wind power, which is by its nature variable. Incorporating AI into regulatory platforms enables governments to shift from the antiquated paradigm of cyclical rule-making to real-time regulatory fine-tuning, one that continuously improves grid resilience, market efficiency and environmental outcomes. In addition, reinforcement learning models can simulate and learn the performance of regulations, optimizing them for subsequent versions. Adaptive governance systems of this nature are critical for achieving renewable energy targets in an ever-changing energy landscape.

7.4.4. *Engaging stakeholders and analyzing public sentiment*

A regulation works well only if there is continuous dialogue with communities, developers and environmental groups. Using AI, and NLP in particular, engagement at scale is made possible through the analysis of public feedback loops (open consultations, social media, local news, etc.) for stakeholder engagement. Sentiment analysis software can help identify local and regional attitudes, fears and support, or resistance levels around wind power projects. These data allow regulators to anticipate needs and develop tailored outreach communication strategies and improved proposals that are more likely to reflect public priorities.

Chatbots and AI-based digital platforms can also be used to enhance two-way interactions enabling the public to ask questions, access personalized information or send feedback in a more streamlined manner. Furthermore, clustering algorithms can organize similar issues or recommendations by stakeholders, assisting policymakers in detecting trends and prioritizing actions. Regulators should look to AI to democratize and scale stakeholder engagement to formulate wind energy policies that are more inclusive, responsive to stakeholder concerns and socially accepted.

7.4.5. *Grid integration and regulatory planning forecasting*

This allows AI to support regulatory decisions around the integration of wind power into the grid, helping steer policy decisions to make grids more agile. Wind energy generation forecasting is sometimes a determinant factor for the stability and planning of electric power grids; thus, short-term and long-term forecasting are crucial. High-precision datasets, such as those from deep learning and ML models trained on meteorological and operational data, accurately predict wind speeds, energy generation and demand surges. These forecasts determine regulations concerning load balancing, energy storage requirements and reserve margins. AI-generated forecasts can be used by regulators to provide dynamic dispatch protocols, update integration thresholds or build flexible grid codes that maximize reliability and minimize curtailment. In addition to these simulated projects, AI can model the interaction of wind farms with other grid resources (such as solar, storage or demand response systems), which will better inform regulators of how to plan hybrid renewable systems. As penetration levels rise, these AI applications guarantee that wind energy can safely and efficiently be incorporated into the grid (Shobeiry 2024).

7.4.6. *EIA optimization*

AI can lead to substantial improvements in EIAs, an important regulatory criterion that must be met for wind energy projects. The conventional EIA process is labor-extensive and dependent upon both manual data collection as well as subjective judgment. The speed and accuracy of assessments are enhanced by AI. ML algorithms can be used to analyze historical environmental data, geographic information systems (GIS) and biodiversity databases to predict potential impacts on wildlife, vegetation, noise levels and visual aesthetics. Through the pairing of remote sensing with computer vision, habitat classification and species detection can be automated in proposed project areas. NLP tools can also analyze existing EIAs to identify recurring themes or precedents that may apply to new proposals. AI can enable regulators to screen high-risk sites early in the process, thus reducing delays and ensuring that wind farms are located in ways that are ecologically responsible. This data layer provides more objective, transparent and consistent environmental reviews, as well as building confidence in the general arenas. Compliance among regulators, the community and business can also be witnessed, and at the same time ecology is protected.

7.4.7. *AI-driven incentive design and subsidy allocation*

Crafting efficient incentive structures for wind energy involves comprehending market trends, cost structures and technological advancements. AI can drill through large datasets: these are data on capital costs, operational performance, investor behavior, market forecasts and other data to optimize subsidy schemes: feed-in tariffs, tax credits, green certificates, etc. Predictive models determine the lowest sustainable incentives to drive development without distorting markets or breaking budgets. Reinforcement learning applicable to all or high-dimensional policy mechanisms in complex systems observed throughout different economic environments allow regulators to simulate countless policies they could offer. This tests which policies can effectively ground their subsidy framework before actively implementing it. AI can also track the impacts of subsidies in real time, enabling adaptive policy calibration as market conditions change. AI is able to make incentives location-specific, technology-specific and profile-specific, leading to more equitable, cost-effective and national energy target-aligned support mechanisms. In fact, AI enables governments to transition from one-size-fits-all subsidies to smart, data-informed incentive models that drive wind adoption while mitigating fiscal risk.

7.4.8. *Harmonization of cross-border regulations*

As wind energy becomes more integrated across geographies and countries, particularly under interconnected grids, cross-border regulatory alignment will become increasingly important. AI can help to map, compare and harmonize the legal frameworks, technical standards and market mechanisms of different jurisdictions. NLP tools can also compare policy documents across multiple languages, flag inconsistencies in approaches and identify opportunities for convergence. Knowledge graphs can help visualize these interdependencies, where regulatory provisions intersect with stakeholder interests, which may be subject to technical requirements across borders. AI can similarly model the implications of differing regulations on cross-border energy flows, market efficiency or environmental outcomes. But this is especially useful for international institutions (such as the EU or regional energy unions) that seek to align grid codes, emission targets, as well as permitting processes. AI-supported data have helped transnational wind energy policymakers to comprehend the obstacles in transnational wind energy generation through their potential benefits, risk factors and legal harmonization.

7.4.9. *Decentralized energy market smart contract regulation*

As blockchain and decentralized energy trading platforms become more common, AI is critical for managing smart contracts – self-executing contracts with the terms of the agreement directly written into code. For instance, smart contracts can facilitate energy sales, curtailment compensation or carbon credit issuance in wind power systems. AI in these contracts can analyze, audit, monitor and optimize them per regulatory standards. Anomaly detection algorithms, for example, can identify fraudulent conduct, misreporting of generation or contractual loopholes. Regulators can also deploy AI to track the effects of new rules on decentralized trading platforms, and how smart contracts behave under stress scenarios. In addition, AI can maintain uniformity and verification of templates of contracts between projects and jurisdictions, consequently decreasing legal ambiguity.

With the potential rise of peer-to-peer energy markets, the combination of AI and smart contract oversight will be key in this space to ensure secure, fair and transparent transactions within a changing regulatory environment (Ohalete et al. 2023).

7.4.10. *Optimization models for land use and spatial planning*

Policy regarding land-use planning is among the most contested, and technically most challenging, aspects of wind power. AI facilitates spatial planning through a multi-layered data overlay to analyze topography, land ownership, environmental sensitivity, infrastructure access and community sentiment. Regulators are using GIS and ML models to find the best places for wind farms to be built, so locations will not conflict with environmental issues and other spatial areas (residential areas) and it can maximize energy production. It can simulate future land-use patterns based on urban growth, crop development and climate change, allowing policies to proactively steer the site of wind expansion.

NLP tools may also extract zoning constraints from municipalities' codes, and predictive models can estimate the effects of policy changes on land values or migration patterns. These insights help policymakers craft zoning laws, buffer zones and planning rules that find a proper balance between energy concerns and prudent development. AI helps to visualize proposed projects in 3D or virtual environments, facilitating community consultations and helping to assuage opposition. AI thereby enables future land-use policy to be inclusive, prescriptive and capable of aligning ecological and energy goals by supporting more intelligent spatial governance.

7.5. Case studies and global best practices

7.5.1. *United States: AI for wind forecasting and grid integration*

AI is increasingly being used in the US, driving improved grid integration and maximized wind energy generation. The US Department of Energy (DOE), along with some national labs and AI companies, have developed advanced wind forecasting models using ML algorithms. These models predict wind patterns and energy generation 48 hours in advance, allowing grid managers and traders to make better decisions regarding energy use and costs. For example, the California Independent System Operator (CAISO) integrates large amounts of wind power with minimal grid disturbance using AI-based predictive analytics. AI-powered systems specify changes from real-time wind generation to help operators adapt energy dispatch strategies, therefore reducing curtailment. Moreover, the US has established regulators that apply AI-enabled decision-making to their permitting processes. Wind farm site selection can be optimized to minimize areas of significant social and environmental impact through automated tools and data-driven modeling. Account-based AI applications are majorly shrinking permitting timelines and enhancing environmental compliance. In doing so, the US is showing how leveraging data-driven technologies that use AI to improve forecasting and regulatory processes can, on the one hand, make wind power a fully connected part of the energy grid and, on the other, make the policy process itself sustainable and efficient.

7.5.2. *Denmark: AI for enhanced wind turbine and energy storage performance*

Recognized as one of the leading countries in wind energy, Denmark has also called on AI by applying the technology to improve wind turbine performance and use energy storage solutions to stabilize the grid. AI-based condition monitoring systems used in a number of Denmark's wind farms analyze real-time data being produced by the turbines and help in strengthening the machine itself through detecting any mechanical failures before they occur. ML models scan vibration, temperature and acoustic data from sensors on wind turbines to identify patterns. Through this predictive maintenance approach, downtime has decreased substantially while the operational efficiency of turbines has improved. Denmark is also using AI in energy storage systems to solve wind power intermittency. For example, the Danish Energy Agency (DEA) is testing using AI to optimize energy storage solutions by automatically determining when to charge and discharge batteries, based on wind power availability, grid demand and market prices. Such real-time optimization allows for the storage of surplus wind energy when generation is high and release during peak demand periods. This not only makes wind power financially viable, but it also assists in stabilizing the grid. Denmark's method is a best practice example to apply AI to both turbine management and wind power storage.

7.5.3. *European Union: AI for harmonizing regulation and cross-border collaboration*

The EU has also, through necessity, engaged in examining the role of AI in harmonizing wind power regulations across member states and enhancing cross-border cooperation. In this sense, through the establishment of a large AI repository, AI is used to harmonize renewable energy policy and regulatory frameworks between EU member states, for the coordination of grid codes and the development of market regulations. The "Clean Energy for All Europeans" package of the European Commission includes AI-driven platforms that facilitate the sharing of regulatory knowledge and best practices among EU member states. AI is optimizing trading and integration of cross-border energy via the European Network of Transmission System Operators for Electricity (ENSTO-E) project. AI tools forecast electricity flows, model market dynamics and optimize cross-border capacity – which allows for the more efficient trading of wind-generated electricity across borders. This helps reduce market inefficiencies and improve energy security through real-time supply and demand balancing. Moreover, the EU's use of AI for regulatory harmonization can be considered a strong global best practitioner, as it promotes cross-border cooperation and enhances cross-border political commitment to the EU's ambitious renewable energy targets. It does so by making the alignment of policies more data-driven and efficient.

7.5.4. *China: application of AI to wind energy site selection and environmental impact assessments*

China is one of the biggest producers of wind energy globally and has explored the use of AI to enhance wind farm siting and EIA. China's National Energy Administration (NEA) is using AI algorithms to find the best location for wind farms by analyzing geological, meteorological and environmental data. Based on the data assigned to these different variables – examples such as wind speed, terrain, distance to grid infrastructure and environmental sensitivity – ML models enable decision-makers to identify locations that will maximize energy production with the least ecological impact. China is incorporating AI into the EIA process, which supplements project site selection with environmental system coding, i.e. an eco-matrix. AI-enabled systems sift through satellite imagery, drone footage and ground data to determine the potential impact of proposed wind projects on local ecosystems, biodiversity and wildlife habitats. In automating the EIA process, China is cutting the timing and expense of manual assessments, streamlining project approval while maintaining environmental integrity. This is critical for balancing the swift development of wind energy with environmental preservation. China's integration of AI into both site selection and EIAs provides a model for scaling wind power infrastructure efficiently and sustainably on a global level.

7.5.5. *India: AI applications in wind power forecasting and regulatory compliance*

India is also turning to AI technologies to reduce the need for human workers and secure regulatory compliance in its decently sized wind power sector. Such AI-based systems have already been put into practice at the National Institute of Wind Energy (NIWE) in India, which serves to predict wind patterns as well as optimize turbine performance at the vast network of wind farms around the country. These AI models use meteorological information in combination with real-operation data from wind turbines to make predictions on wind speeds, energy production and curtailment events. In this way, grid operators can distribute energy efficiently and avoid wasting power generated by surpluses in the wind. Use of AI in streamlining the regulatory compliance environmental monitoring and reporting is one area where there are serious regulatory compliance issues and hence AI is also being embraced. Automated AI tools monitor noise, emissions and land use around wind projects to make sure they follow local commitments. AI drones and satellite imagery for environmental monitoring provide real-time data on land degradation, wildlife impacts and ecosystem changes around wind farms. With AI harnessed for operational optimization as well as regulatory oversight, India is setting an example for nations with ambitious renewable energy targets.

7.6. Future direction

7.6.1. *Integration of AI with smart grids for decentralized energy systems*

The deep integration of AI with smart grids represents a critical frontier in the future of wind power regulation. With higher power generation from wind energy, grid stability will be implemented through connections with decentralized systems. In contrast, AI can facilitate decision-making in real time at the edge of the grid by adjusting energy distribution to synchronize demand, supply and grid conditions. The proliferation of distributed energy resources (DERs), from small wind turbines to battery storage systems, means that when these increasingly diverse resources are connected to microgrids and virtual power plants (VPPs), AI is needed to manage energy flows. Future AI solutions will not only monitor and control individual wind turbines but optimize the entire energy ecosystem, including demand response, energy storage and solar energy generation. AI will help grid operators to adapt dynamically for the intermittency of wind power and other renewables, integrating them and minimizing peaking power of fossil fuels. Futuristic smart grids powered by AI will, therefore, lead the way for more flexible, efficient and robust energy systems capable of adapting to changing grid conditions more dynamically, while maximizing the share of renewables on the grid (Danish and Senjyu 2023b).

7.6.2. *Powerful AI for modeling policy predictive modeling*

In the future, AI will allow for far more sophisticated predictive modeling for wind power policy. Policymakers will make decisions based on AI-driven simulations not just by evaluating the effects of existing regulations, but by predicting the effects of potential future policy interventions. In particular, reinforcement learning algorithms will be used to rapidly simulate competing policy scenarios in real time and deliver insight on optimal regulatory pathways that enable accelerated wind power deployment while simultaneously attaining broader energy, economic and environmental objectives. Such models will incorporate a range of dynamic inputs including technological progression, market response, political shifts and climate fluctuations to deliver more precise and responsive policy recommendations. AI will enable policymakers to produce future-proof regulations far more rapidly and respond at a quicker pace to changing technological and market developments, with the long-term goal of hastening the global transition to renewable energy by bolstering policy flexibility and responsiveness (Mitchell et al. 2022).

7.6.3. *AI for fostering cross-sector collaboration*

With wind regulation becoming more complicated, AI will enable cooperation between sectors. The AI systems of tomorrow will be designed to help regulators, developers, environmental groups and communities work together more productively by creating shared and transparent platforms for making decisions. AI tools will be harnessed to integrate and link these data across the energy, transport, agriculture and urban planning sectors, helping to ensure that wind energy projects are aligned across sectors and that wind policies are integrated into wider national agendas for sustainable development. AI models, for example, could corral wind energy with EV adoption, determining how integrating more EVs into the grid may alter when wind power is "needed" and how much must be stored for use. AI will facilitate the creation of cohesive energy ecosystems centered around sustainability and inter-dependencies between energy, transportation and land use through policies across multiple sectors. This will go a long way to helping address some of the biggest challenges we face in the world today, among them, the drive towards net-zero emissions and the challenges of delivering on climate commitments (Ohalete et al. 2023).

7.6.4. *Transparent and inclusive decision-making by AI*

This will bring AI technologies closer to the public, allowing communities to participate in the policymaking process more successfully. NLP will allow stakeholders to dive into complex regulatory documents and provide comments, thoughts and/or alternative policy options. AI chatbots and virtual assistants could help the public navigate regulatory processes, answer frequently asked questions and make recommendations based on user inputs. Furthermore, AI will enable equitable policymaking through real-time analysis of social and environmental externalities. By analyzing how policies affect different demographics and regions, AI tools can pinpoint inequities in access to wind energy and assist regulators in crafting more equitable solutions. This will be increasingly relevant as countries expand their generation of wind power by deploying infrastructure into far-flung, marginalized or biodiversity-rich spaces, where, in particular, local communities should be properly consulted and compensated for their inclusion in wind power expansion (Shobeiry 2024).

7.6.5. *The rise of quantum computing in wind power optimization*

As quantum computing becomes more mature in real-world terrestrial settings, its synergy with AI could greatly optimize wind energy production. Quantum computers can tackle extremely complicated problems much faster than classical computers and are ideal for tasks such as modeling wind patterns, optimizing turbine design or simulating large grid operations. For example, quantum algorithms and AI can significantly improve turbine layout optimization in order to minimize wind farm costs while maximizing energy capture. Quantum computing can also enhance energy systems in several ways, such as improving the chemistry involved in energy storage through better battery chemistry simulations or optimizing the way energy storage is managed for wind energy so that large volumes of energy produced during hours of maximum wind production can be stored. Considering quantum computing is still in its infancy, its eventual fusion with AI will enable the wind energy field to have capabilities never seen before, increasing efficiency and performance while decreasing price.

7.6.6. *Blockchain and AI for new energy trading and grid decentralization future directions*

AI and blockchain will be able to ensure secure and transparent energy trading. For example, integrating AI-based optimization with an immutable blockchain ledger can provide a trust factor for wind power trading in the decentralized energy market. AI will predict electricity prices and optimize wind power trading between producers, consumers and grid operators. Blockchain will make sure these transactions are completely transparent, traceable and not easy to alter. AI-enabled blockchain networks will also facilitate peer-to-peer energy trading, where small-scale producers (e.g. homeowners with small wind turbines) can sell excess energy directly to consumers. AI algorithms will be used to optimize trading strategies, aggregate transaction costs and keep prices fair in this decentralized landscape. Data-driven AI will again play a crucial role in forming the affordable portfolio of the Internet of Energy, enabling new energy customers to become active participants in their energy exchange through decentralized microgrids – known as prosumers (producer–consumer of energy) – that will contribute to a new energy paradigm, and that will also impact the increase in wind power stakeholders. Finally, AI technologies will be essential to increasing the resilience of wind power systems as climate change continues to impact weather patterns. AI will also assist in modeling the effect of extreme weather events, from hurricanes to floods to droughts, on wind farms and for recommending mitigation methods. Using computer climate change models, AI systems will be used to analyze millions of bytes of climate data to predict how the availability of wind resources will change over time based on shifting climate conditions. Incorporating climate projections into regulations around new wind power generation is an example of how AI can assist policymakers in designing infrastructure that is more resilient to increasingly severe weather events. Moreover, AI will use climate-adaptive technologies, including wind turbines, specifically engineered to function under extreme conditions, or even automated systems that dynamically change turbine configurations according to live updates from weather forecasting models. This will ensure that wind power continues to be a dependable energy source as the climate changes.

7.7. Conclusion

To conclude, AI can cultivate a stronger collaboration between wind stakeholders by increasing the integration of information, optimizing management processes and harmonizing regulatory processes. ML, predictive analytics and optimization algorithms are some of the AI technologies that can allow for more accurate forecasting of wind and improved performance management of turbines and grid integration. With the power of AI, regulators can also improve permitting procedures, EIA and incentive schemes. Furthermore, the use of AI enables the creation of dynamic, real-time versions of policy models that adjust in response to market shifts and technological leaks, rendering wind power policy proactive and future-centric. Looking ahead, the integration of AI with advancements such as smart grids, energy storage optimization and quantum computing holds the potential to elevate wind power systems to new heights of flexibility and resilience, so that they respond better to the dynamic challenges posed by climate change and variations in energy demand. Governments and industry leaders should use AI to promote the transition to wind energy, thereby helping to build renewable energy that is sustainable and resilient as energy transitions accelerate around the world. This will also enhance the way wind energy operates, streamlining the process from analysis to planning and action – leading to overarching efficiencies and profitabilities with the end goal of a low carbon society – with greater energy security.

7.8. References

Bošnjaković, M., Martinović, M., Dokić, K. (2025). Application of artificial intelligence in wind power systems, *Applied Sciences*, 15(5), 2443.

Chahar, S., Gupta, S., Dhingra, I., Kaswan, K.S. (2024). Adversarial threats in machine learning: A critical analysis. In *International Conference on Computational Intelligence and Computing Applications (ICCICA)*, Samalkha, 253–258.

Chahar, S., Kaur, K., Kaswan, K.S., Malik, K. (2025). Intelligent control mechanisms for renewable energy integration using machine learning models. In *International Conference on Pervasive Computational Technologies (ICPCT)*, Greater Noida, 719–724. IEEE.

Danish, M.S.S. and Senjyu, T. (2023a). AI-enabled energy policy for a sustainable future. *Sustainability*, 15(9), 7643.

Danish, M.S.S. and Senjyu, T. (2023b). Shaping the future of sustainable energy through AI-enabled circular economy policies. *Circular Economy*, 2(2), 100040.

Lüthi, S. and Prässler, T. (2011). Analyzing policy support instruments and regulatory risk factors for wind energy deployment – A developers' perspective. *Energy Policy*, 39(9), 4876–4892.

Mitchell, D., Blanche, J., Harper, S., Lim, T., Gupta, R., Zaki, O., Tang, W., Robu, V., Watson, S., Flynn, D. (2022). A review: Challenges and opportunities for artificial intelligence and robotics in the offshore wind sector. *Energy and AI*, 8, 100146.

Ohalete, N.C., Aderibigbe, A.O., Ani, E.C., Ohenhen, P.E., Daraojimba, D.O., Odulaja, B.A. (2023). AI-driven solutions in renewable energy: A review of data science applications in solar and wind energy optimization. *World Journal of Advanced Research and Reviews*, 20(3), 401–417.

Shobeiry, S.M. (2024), AI-enabled modern power systems: Challenges, solutions, and recommendations. In *Artificial Intelligence in the Operation and Control of Digitalized Power Systems*, Azad, S. and Nazari-Heris. M. (eds). Springer, Cham, 19–67.

8

Implementation of an AI-Driven Wind Energy Sustainability Framework Using Reinforcement Learning-Optimized Deep Neuro-Fuzzy Controller (RL-DNFC)

This chapter presents the design of an AI-based system of wind power sustainability based on a Deep Neuro-Fuzzy Controller (DNFC), which functions through reinforcement learning. The proposed system is the combination of deep neural learning, the rationale-based on fuzzy logic and the maximization of the reinforcement learning in order to achieve adaptive and sustainable control of the wind turbines. DNFC possesses interpretability and nonlinear mapping capabilities, and the reinforcement learning layer, which is trained using the Deep Deterministic Policy Gradient (DDPG) algorithm, is constantly optimized to provide the best possible performance. The model was written in Python on the TensorFlow platform and trained on the NREL wind data under various wind speeds. The experiment results indicated that the accuracy of power output has increased substantially with an RMSE of 0.27 kW and a power coefficient (Cp) of 0.47. The system reduced by 23% of the torque variation and increased the life of turbines by approximately 18.7. The findings support the argument that the RL-DNFC scheme is an effective method to maximize power usage, flexibility and output, developing an intelligent solution based on scalability to the future generation of wind energy systems.

8.1. Introduction

Renewable energy sources such as wind power have become one of the most promising sources of clean electricity in large volumes due to the shift toward more

Chapter written by Rajasekaran PALANIAPPAN, E. Afreen BANU, P. ROBERT and J.D. Dorathi JAYASEELI.

sustainable energy systems. The efficiency and reliability of wind energy systems are, however, often limited by very nonlinear and stochastic conditions of the environment. The perturbation of wind, the fast change of velocity and the difference in the air density add uncertainty to the performance of turbines; hence, the control of pitch angle, direction of yaw and the generator torque is a complicated adaptive issue (Sundaresan et al. 2023). The traditional control systems such as Proportional–Integral–Derivative (PID) and non-adaptive fuzzy controllers have little flexibility in dynamically varying wind patterns and generate sub-optimal energy harvesting and mechanical loading (Bhowon et al. 2022).

Adaptive learning and self-optimization have become new techniques of researching such as nonlinear dynamics using artificial intelligence (AI). The neural networks, namely the convolutional neural network (CNN) and recurrent neural network (RNN), have proven to be successful in wind speed prediction and power curve estimation (Liu and Zhang 2024).

On the same note, fuzzy logic systems are also associated with interpretability and strong decision-making in times of uncertainty. However, by combining deep neural learning and fuzzy inference, the system is able to learn more complicated input–output maps and maintain rule-based reasoning that can be understood by people. The result of this combination is the Deep Neuro-Fuzzy Controller (DNFC), which can learn as well as reason (Salameh et al. 2025).

Amid these benefits, adjustment of the controller parameters online to fit real-time variations in the environment is one of the issues that have been persistently challenging. Reinforcement learning (RL) is a subdivision of AI based on behavioral psychology, which provides an effective approach to autonomous adaptation by engaging the environment. The RL agents determine the best control policies that maximize cumulative rewards in the long term by defining the states, actions and reward functions (Wu et al. 2023). Combining RL and DNFC leads to a Reinforcement Learning-Optimized Deep Neuro-Fuzzy Controller (RL-DNFC)-hybrid intelligent system that can continually learn and improve itself and be robust against wind profile changes (Liu et al. 2025).

This chapter aims to develop and apply a wind energy sustainability model with AI-powered technology, which uses the RL-DNFC model to optimally control the working of wind turbines. The proposed structure will optimize the energy conversion efficiency, reduce mechanical loads and be sustainable to operate in different environments. It is implemented with Python (TensorFlow/PyTorch) to integrate deep learning with RL on the basis of real-world wind datasets obtained via the National Renewable Energy Laboratory (NREL). The performance of the system is determined through energy sustainability variables, which include power coefficient (C[?]), output power and load smoothness, and AI performance metrics, which include root-mean-square error (RMSE), mean absolute error (MAE) and reward convergence.

The remaining sections of the chapter are structured in the following way: section 8.2 will give a literature background on AI-based wind energy optimization methods, section 8.3 will discuss the architecture of the proposed system, section 8.4 will give the methodology and mathematical formulations of the RL-DNFC algorithm, section 8.5 will describe the environment and parameters of the simulation, section 8.6 will give the experimental results with detailed tables and graphs, section 8.7 will discuss and comparatively analyze and section 8.8 will give the conclusion of the chapter with the key.

8.2. Literature review

The development of the wind energy systems has shifted away from fixed control systems to intelligent and adaptive systems that can learn by themselves and can optimize their operations in real time. The literature around this field is widely diverse and covers three dimensions of technology fuzzy logic control (FLC), neural and deep learning system and RL-based optimization. Both of these methods have their own advantages as well as limitations that are inherent when implemented individually. Therefore, recent studies have aimed at the hybridization of these paradigms to increase the energy sustainability, operational and environmental flexibility.

8.3. Wind energy control using fuzzy logic

The use of FLC in wind turbine management has been an acclaimed concept owing to the fact that the technology can be used to represent linguistic uncertainties of the variation in wind speed and weather conditions (Kumar et al. 2025). A fuzzy system will be able to represent nonlinear relationships without the need for precise mathematical representation as in the case of traditional linear controllers.

Indicatively, Simoes and colleagues (Hamoodi et al. 2020) designed a fuzzy-based pitch angle controller, which produced a more smooth power curve when the wind was gusty. The rule base and membership functions used in classical FLCs are, however, normally developed in a heuristic manner, resulting in poor adaptability to environmental conditions that do not conform to the training domain.

Later, fuzzy systems based on hybrid Fuzzy-PID and Adaptive Systems were introduced to make them responsive (Kusiak and Verma 2010). These systems were dynamically tuned to fuzzy parameters based on heuristic tuning, but not based on data-driven optimization. Therefore, even though FLC was offering a strong initial control, its capability to operate in stochastic and high-dimensional wind conditions was still not optimal, which is why the idea of learning-based systems was welcome.

8.4. Neural and deep learning models to predict wind power

ANNs have emerged as a leading instrument in the forecasting of wind energy mostly because of their capability to estimate the complex nonlinear mappings of environmental variables to the output of turbines. Initial studies used multi-layer perceptrons (MLP) and radial basis function (RBF) in the prediction of wind speed and energy production (Benti et al. 2023). CNNs and long short-term memory (LSTM) networks have since become the deep learning architectures used to learn the spatio-temporal patterns in wind data (Kumar et al. 2023). Indicatively, Zhang and colleagues (Alhussein et al. 2020) suggested a CNN-LSTM hybrid model of short-term wind power prediction, which had a mean absolute percentage error (MAPE) of 6.24% on the NREL database. Multivariate meteorological data has also been analyzed through deep autoencoders to extract hidden representations that allow more accurate prediction of the power curves (Kosana et al. 2021). The majority of deep learning models, however, are black boxes that are not interpretable and physically consistent, among other factors which are essential in making decisions about the control of turbines. The problem of uncertainty management and interpretability through rules of deep models triggered the interest of researchers in understanding Neuro-Fuzzy Systems (NFS). An Adaptive Neuro-Fuzzy Inference System (ANFIS) is a neural learning model that integrates the fuzzy reasoning and, as a consequence, allows the neural learning model to learn patterns, as well as make decisions based on rules (Jang 1993). However, the non-adaptive ANFIS models do not easily change accordingly in real time to changes in the environment during the operation of the turbine.

8.5. Adaptive wind energy control with reinforcement learning

RL is a paradigm shift of supervised prediction to autonomous control. In contrast to predictive models, which use pre-labeled data, RL agents are able to optimize control policies by interacting with the environment via trial-and-error processes (Sutton and Barto 2021). RL has also been used to maximize cumulative rewards in wind turbine systems (maximizing energy output and penalizing mechanical stress) by maximizing pitch angle adjustment and yaw adjustment.

Indicatively, Li and colleagues (Soler et al. 2024) applied an adaptive controller designed with Q-learning to optimize pitch angle tuning to gain 5.6% improvement in the mean power output over PID control. Deep Deterministic Policy Gradient (DDPG) and Proximal Policy Optimization (PPO) are deep reinforcement learning techniques that contributed to the increased accuracy of control when operating in continuous state-action spaces (Zholtayev et al. 2024). Nevertheless, RL models generally need a large number of training episodes to reach steady convergence and are not usually interpretable without a reasoning layer, for example, fuzzy inference.

8.6. Neuro-fuzzy and hybrid reinforcement approaches

Method	Core principle	Limitations	Performance metrics	Reference
Fuzzy logic control (FLC)	Rule-based decision using linguistic variables	Static membership tuning; non-adaptive	Power fluctuation ± 8%	(Hamoodi et al. 2020)
ANN-based prediction	Data-driven pattern learning	Black box; lacks interpretability	RMSE = 0.45 kW	(Benti et al. 2023)
Deep LSTM model	Spatio-temporal learning	No reasoning capability	MAE = 0.36 kW	(Alhussein et al. 2020)
Q-learning control	Reward-based optimization	Discrete state limitation	Power gain + 5.6%	(Soler et al. 2024)
Deep RL (DDPG/PPO)	Continuous action optimization	Convergence instability	RMSE = 0.31 kW	(Zholtayev et al. 2024)
ANFIS	Hybrid neural-fuzzy reasoning	Static structure	MAE = 0.42 kW	(Jang 1993)
Fuzzy-DRL hybrid	RL-tuned fuzzy parameters	Computationally expensive	Power gain + 7.8%	(Khooban and Gheisarnejad 2020)
Proposed RL-DNFC	Deep neuro-fuzzy controller with RL optimization	–	RMSE = 0.27 kW; power gain + 10.3%	Proposed

Table 8.1. *Comparative review of AI techniques in wind energy control systems*

Hybrid models have become a better alternative to overcoming the idiosyncrasy of deep learning, fuzzy inference models and RL models. These systems combine learning ability, transparency of reasoning and real-time flexibility. As an example, Chen and colleagues (Khooban and Gheisarnejad 2020) created a Fuzzy Deep Reinforcement Learning (FDRL) system which dynamically adjusted membership functions using RL feedback, enhancing power stability by 7.8% over time. In the same light, Panwar and colleagues (Nguyen et al. 2025) proposed a Deep Neuro-Fuzzy Reinforcement Controller (DNFRC) of loading in microgrids, which showed superior convergence and oscillation decreases. According to the reviewed literature, there is a progressive shift toward that of rule-based fuzzy systems to deep learning and hybrid RL systems. Not one of the available frameworks, however, integrates adaptive control, interpretability and sustainability goals holistically. Table 8.1 provides a summary of comparative features of current methodology and outlines the novelty of the offered RL-DNFC approach.

Although there has been this kind of progress, not many studies have carried out an RL-DNFC in the context of wind energy sustainability. The majority of the hybrid models are either simulated with no real dataset verification or only aim at optimization of performance, but do not consider the sustainability factor (e.g. mechanical load reduction and turbine life increase). This study fills this important gap by using a complete RL-DNFC model that balances between maximizing energy and sustainability indicators.

8.7. System architecture of the AI-driven wind energy sustainability framework

The proposed AI-based wind energy sustainability system will be programmed to create an intelligent, adaptive and self-educating control system that will streamline the performance of the turbines under dynamically evolving environmental conditions. The goal of the framework is to maximize the efficiency of the energy conversion and increase the reliability of the system and its sustainability in the long term by incorporating deep learning, fuzzy reasoning and RL into the single decision-making space. The architecture is a closed-loop system, which has interrelationships between data collection, learning, optimization and actuation based on continuous feedback. This setup allows the model to change its control strategies dynamically according to its real-time performance of the turbines and the environment.

On the basic level, the framework functions on five interrelated layers, which include the data acquisition layer, preprocessing and feature extraction layer, deep neuro-fuzzy control layer, RL optimization layer and actuator control layer. These layers are a combination of raw environmental data into optimized control actions.

The architecture is hierarchical in that low-level modules are involved in sensing and data processing, whereas high-level layers are involved in decision-making and adaptive control optimization. The architecture will ensure a smooth connection between the physical world and the AI learning core to create a cyber-physical loop of sustainable wind energy management.

The process starts with the Wind Data Acquisition Unit (WDAU), which is the sensory part of the system. There are critical environmental and operational parameters that are collected by this unit and include wind velocity, wind direction, air density, rotor angular speed and power output. The information is recorded using sensors using the IoT and sent in real time to the processing unit.

Wind behavior is stochastic and it tends to be noisy; hence, the raw data normally contains irregularities that have to be filtered first before being used to make control decisions. This crude flow of sensory information therefore forms the basis on which processes of intelligent decision-making are founded.

After being acquired, the data is added to the Data Preprocessing and Feature Extraction Module (DPFEM). This module is important in making sure the information involved is accurate and reliable to the AI components. It carries out noise filtering, normalization and feature computation processing steps to remove redundancies and stabilize the input values. Low-pass filtering eliminates high-frequency noise of sensors, and the input features are normalized to a standard numerical range in order to ensure that the neural training converges more quickly. Statistical descriptors that are also calculated with the module include mean wind velocity, standard deviation of turbulence and moving average of energy output. The characteristics represent a representative input vector which describes the present state of the turbine and its environment.

After the input features are created, they are presented to the DNFC which is the heart of the decision-making unit of the architecture. DNFC combines interpretability of fuzzy inference systems and nonlinear learning of deep neural networks. The controller converts the processed inputs to control commands to the turbine, i.e. the best pitch angle, generator torque and yaw control. The fuzzification layer of the DNFC transforms the numerical features to linguistic variables through Gaussian membership functions. The variables are then subjected to a rule inference mechanism that is used to model expert knowledge as a set of rules that have an if-then form. An example is that, in case of high wind speed and a moderate rotor speed, the pitch angle should be adjusted slightly downward to avoid overloading. These fuzzy parameters are refined by the neural aspect of the DNFC in a manner that is responsive to variations in operating patterns by dynamically changing the membership centers and the weights of the rules. The DNFC output is a combination of sharp control signals that are commensurate with the prevailing optimal turbine setup.

The DNFC produced decision outputs that are further optimized by the Reinforcement Learning Optimization Layer (RLOL). This stratum brings in flexibility as the performance of the DNFC is constantly checked and its internal parameters change in response to the environment. The optimization is described as Markov Decision Process (MDP), with every system state representing a combination of wind and turbine states and the action as certain control manipulations, which may change in pitch or torque. RL agent measures the state, takes an action and gets a reward which is a positive measure of the action taken. The reward function is aimed at the maximization of the long-term energy output and minimization of stress and mechanical wear. Based on repeated policy changes, the RL agent gets to develop strategies on how to pick the most appropriate control action in every particular environmental condition and optimizes the efficiency of the system with time. This self-learning will enable the system to perform highly even during the unpredictable weather conditions, hence improving operational resilience and sustainability.

Optimized control commands are then sent to the Decision and Actuator Control Unit (DACU) which is the physical interface between the AI controller and the turbine hardware. This unit reads out the control signals of the RL-DNFC and makes related mechanical changes. The DACU controls three main control parameters: the pitch angle which adjusts the aerodynamic lift of the blades; the yaw orientation which ensures that the rotor is facing the wind direction; and the generator torque which balances the mechanical load and the electrical output. The RL agent receives continuous feedback of the actual response of these actuators sent by the sensors. This feedback mechanism helps the system to assess whether the actions performed were of the intended effects and the learning cycle is closed and the system is able to continue adapting.

There is a communication layer that facilitates the communication of all modules that ensures reliability in the transfer and synchronization of data. The architecture makes use of the Internet of Things (IoT) to monitor and control remotely. The turbine sensors and the AI engine communicate in real time by using lightweight messaging protocols such as MQTT, which allows transmitting in real time and scaling to several turbines. Both the edge computing and cloud integration are supported by the data infrastructure: edge devices can execute time-sensitive inference to act quickly, and cloud servers are capable of training models in the long term and updating policies. This hybrid design of computing should provide a balance between real-time responsiveness and the efficiency of computing, which is essential to ensure the sustainability of the large-scale wind energy farms.

The inner data flow of the system is based on an endless feedback. First, there is a collection and processing of environmental and turbine data to create a state vector to describe the operating state. The DNFC uses this vector to approximate initial control actions, which are then optimized with the use of RL. The results of the optimized actions are realized with the help of the actuator interface, and the performance indicators (energy output, vibration intensity and load distribution) are observed. These are fed back into the learning module which changes its parameter to reduce the difference between desired and actual results. This information cycle will guarantee that the control system is continuously improved by time and is able to adapt to the dynamic characteristics of wind to sustain operations in the long run without any outside management.

The architecture design consequently represents the qualities of smartness, versatility and longevity. The intelligence is obtained by combining deep neural with the fuzzy reasoning elements that are capable of capturing the dynamics of the complex systems. The RL module gives the framework adaptability as it allows the framework to rectify itself and increase its decision-making with time. By the use of energy-conscious optimization and stress-reducing control measures, which ensure long life cycle of turbines as well as efficiency, sustainability is achieved. All of these features combine to make the proposed system a novel and viable solution to the practical management of wind energy in the real world.

AI-driven wind energy sustainability framework system architecture is a coherent combination of sensing, learning and actuation. The framework overcomes the constraints of the traditional static control system by incorporating the RL-DNFC into a continuous feedback system. It offers a smart system of performance improvement in a dynamic manner that guarantees a high degree of energy and long-term functionality. This architectural background preconditions the methodological implementation and development of algorithms that are discussed in the next section.

8.8. Methodology and algorithmic design

The approach followed in this study to apply the RL-DNFC is designed in a way that it combines the adaptive intelligence and the physical interpretability. The

RL-DNFC model will be the decision core of the proposed AI-based sustainability framework in the field of wind energy. The methodology summarizes four closely interrelated computational processes, namely input feature modeling, deep neuro-fuzzy inference, RL-based optimization and real-time adaptation of control. A combination of all of these processes creates a closed feedback process that is able to learn the best control strategies to balance the maximization of power with the stability of the system with dynamically changing wind conditions.

The first basis of the methodology is the modeling of wind turbine dynamics. The instantaneous mechanical power generated by a wind turbine can be expressed as

$$P_m = \frac{1}{2}\rho A C_p(\lambda, \beta) V_w^3 \quad [8.1]$$

where ρ is the air density, $A = \pi R^2$ represents the rotor swept area, C_p denotes the power coefficient, λ is the tip-speed ratio, and β is the blade pitch angle. The primary control objectives are to maintain the power coefficient near its optimal value $C_{p,opt}$, while minimizing mechanical stress on the blades and generator. The controller must therefore adaptively tune β and generator torque τ based on the prevailing wind speed V_w, air density ρ and rotor speed ω_r. These parameters are not linear and constant so that a traditional linear controller is not able to keep its efficiency on the entire range of operation. Therefore, there is a need to have a hybrid intelligent control model. The first layer of the intelligence is the DNFC. It combines the fuzzy inference rule-based reasoning with a nonlinear approximation power of deep neural networks. The fuzzification stage converts the input features, including wind velocity, direction, air density and the rotor speed, into linguistic variables defined by the Gaussian membership functions. These membership functions are defined as

$$\mu_i(x) = \exp(-\frac{(x-c_i)^2}{2\sigma_i^2}) \quad [8.2]$$

where c_i and σ_i represent the center and spread of the ith membership function, respectively. The fuzzy rule base then combines these memberships using logical operators to infer an intermediate fuzzy output. Each fuzzy rule R_j takes the form: if the wind speed is A_i and the rotor speed is B_k, then the pitch angle is C_l.

The neural layers following the fuzzy inference system perform adaptive tuning of the parameters c_i, σ_i, and consequent weights w_j using gradient-based learning. The defuzzification process finally converts the fuzzy output into a crisp control signal that determines actuator settings for pitch and torque.

The next layer of optimization is governed by RL, which introduces a self-learning ability to the DNFC structure. The problem is formulated as an MDP, characterized by a state space $S_t = \{V_w, \omega_r, \beta, P_{out}\}$, an action space $A_t = \{\Delta\beta, \Delta\tau\}$ and a reward function R_t.

The RL agent interacts with the environment through iterative trials, updating its policy based on the received rewards. The agent's objective is to maximize the cumulative discounted reward defined by

$$J(\pi) = \mathbb{E}[\sum_{t=0}^{T} \gamma^t R_t] \quad [8.3]$$

where $\gamma \in [0,1]$ is the discount factor determining the relative importance of immediate versus future rewards. The reward function is a weighted composite that simultaneously accounts for energy output efficiency and mechanical safety, expressed as

$$R_t = \alpha(\frac{P_t}{P_{rated}}) - \beta(\frac{(\omega_r - \omega_{opt})^2}{\omega_{opt}^2}) - \delta(\frac{|\tau - \tau_{lim}|}{\tau_{lim}}) \quad [8.4]$$

where α, β, δ are the weighting coefficients, P_t is the instantaneous power output, P_{rated} is the rated turbine power and τ_{lim} denotes the torque limit. The former term encourages more power production, the second one punishes extreme speed variations of the rotor and the third one disapproves of torque overcharge, which guarantees the sustainability of the long term.

The reinforcement learner of the RL-DNFC architecture applies the DDPG algorithm, which is well applied to continuous control actions, including pitch and control of torques. DDPG uses two deep neural networks, an actor network that chooses the actions using the existing states and a critic network that measures the quality of the actions by the Q-value function. The Bellman equation is used to update the critic as follows:

$$L(\theta^Q) = \mathbb{E}_{(s_t, a_t, r_t, s_{t+1})}[(Q(s_t, a_t \mid \theta^Q) - y_t)^2] \quad [8.5]$$

where

$$y_t = r_t + \gamma Q'(s_{t+1}, \mu'(s_{t+1} \mid \theta^{\mu'}) \mid \theta^{Q'}) \quad [8.6]$$

and Q' and μ' represent the target networks used to stabilize training. The actor network parameters are updated using the deterministic policy gradient:

$$\nabla_{\theta^\mu} J = \mathbb{E}_{s_t}[\nabla_a Q(s, a \mid \theta^Q) |_{a=\mu(s)} \nabla_{\theta^\mu} \mu(s \mid \theta^\mu)] \quad [8.7]$$

This topology of two networks ensures a stable convergence and gradual policy evolution process in such a way that the control behavior is not oscillating during turbulent wind. The RL and DNFC are united in terms of sharing parameters and linked through feedback. The fuzzy membership parameters and rule consequents under the DNFC are then narrowed down periodically based on the gradients computed based on the RL reward feedback. With this kind of connection, the controller is capable of trading off short-term learning based on neural updates on the data with long-term policy improvement on the foundation of RL. The combined control signal at each time step is computed as

$$A_t = \eta A_{DNFC}(t) + (1-\eta)A_{RL}(t) \quad [8.8]$$

where η is an adaptive blending factor determined by the reward stability index. This ensures a smooth transition between supervised inference-based actions and RL-driven policy adjustments. As the RL component converges, the contribution of $A_{RL}(t)$ increases, leading to progressively autonomous control.

The RL-DNFC model is algorithmically designed in a linear way. The DNFC is first pre-trained on historical data of turbine operation in order to create a stable background control. The RL agent then replaces the deployment in the live environment, and increasingly the control policy is updated by working with it. The general operations include function, experience gathering, reward assessment, backpropagation and updating the network. Reward stability and minimum loss of Q-value are the monitored aspects to monitor the policy convergence. The system stops the training after the performance measures achieve the steady state within a specified tolerance.

To execute it, the TensorFlow platform of Python is used to build the DNFC and the RL agent networks. The DNFC has a four-layer structure with an input layer, two hidden layers with 64 neurons each with ReLU activation and an output layer, which is a representation of control actions. The RL module contains an actor network and a critic network (both of depth three), both optimized with Adam (learning rate 10 -4). The training serves a replay buffer of size 10 5 to store more recent experiences and the batch size of 64 to make stochastic gradient descent updates. The stability of the training is attained with the soft updates to the target networks with the coefficient t = 0.001. All of the structure works in the episodic training mode, whereby each episode is associated with a ten-minute simulated wind-in-real-time situation.

The convergence of the RL-DNFC model is evaluated based on the trajectory of the reward, and convergence of the error and the stability of the control outputs.

These decisions of the controller change dynamically as the agent becomes proficient in keeping the optimal pitch and torque in response to different levels of turbulence. With time, the system reduces variation in errors of energy production and is also mechanically sound. Therefore, the methodological design finds a strong base to the sustainability of energy optimization by means of smart learning mechanisms.

8.9. Implementation setup and simulation environment

The proposed RL-DNFC was implemented in a well-calibrated computational and simulation environment to guarantee realism, reproducibility and computational efficiency. The experimental system combines hardware-level simulation data as well as software-based AI functionalities to simulate the real dynamics in wind energy. The general goal of the implementation stage is to transform the theoretical background presented in the previous sections to a real-life and workable model that is capable of controlling energy in an adaptive and sustainable manner.

The simulation tests were performed using a workstation with an Intel core i9 processor that has 32 GB of RAM and an NVIDIA RTX 4080 with CUDA to support accelerated tensor computations. The programming platform was Python 3.10, and the software used was built on the deep learning and RL development frameworks of TensorFlow 2.14 and PyTorch 2.2. Data handling, numerical analysis and visualization were done using supporting libraries such as NumPy, Pandas, Matplotlib and SciPy. The OpenAI Gym interface was used to construct RL environments with the characteristics of modular policy training and real-time state-action-reward updates. A hybrid Python-MATLAB coupling with MATLAB Engine API to Python was used to simulate aerodynamic and mechanical dynamics of the wind turbine, to give good turbine response modeling, based on physical equations of motion.

The training and evaluation data of the RL-DNFC model was obtained based on the NREL Wind Integration Dataset (WIDS) that offers high-resolution time-series information of wind speed, direction and power output of various turbine locations. Each simulation cycle sampled the data at one second and a period of 24 hours, which was appropriate to provide enough fine detail to track the fast changes in the wind conditions.

To further test the robustness of the algorithm, synthetic data of turbulence were created based on the Kaimal spectral model to model gusty conditions and time-dependent flow patterns. The dataset was broken down into three categories 70% training, 15% validation and 15% testing. All of the continuous input parameters were normalized with the min–max scaling of 0–1 to allow all of the learning algorithms to converge to the same point.

This has a two-phase implementation plan, which includes the pre-training phase and the online learning phase. During the pre-training stage, the DNFC was trained by supervised output based on historical turbine data to estimate the relationship between environmental feature input and optimal control outputs. At this stage, the DNFC membership values and weights of its rules were started and refined based on the values of backpropagation with gradient descent and adaptative learning rate, 10(-3). The training process was extended to 200 epochs until convergence was reached, which was shown by the plateau of the loss function. The RL agent was then added to the model, which constituted the entire RL-DNFC architecture of real-time adaptive control.

The online learning phase was the dynamic part of the system as the RL agent was going through the simulated turbine environment. The reinforcement learning optimization layer used the DDPG algorithm in that it is capable of continuous action spaces. The actor network adopted had three hidden layers of 128, 64 and 32 neurons, respectively, which were activated by the ReLU function and the output layer was by the tanh activation function to generate normalized control actions to pitch and torque adjustment. The critic network was of similar structure but had two streams of inputs to handle both state and action variables and then calculate the Q-value. The critic and actor learning rates were set to 10 -4 and 10 -3, respectively, and the replay buffer size was set at 100,000 experience tuples. The target networks soft update rate was fixed to t = 0.001 in order to make the networks stable and avoid the divergence that could occur in policy learning.

The design of the reward function was important to the attainment of sustainable optimization. The combination of power efficiency, load smoothness and mechanical stability goals was put into a single scalar feedback (the reinforcement signal). At every timestep of the simulation, the reward was determined by the ratio of the actual power production to rated power which was punished by the variation in the rotor speed and torque stress. This made sure that the RL agent does not only learn how to maximize production of energy but also ensure reliability in its operations and increase the life cycle of the equipment. The cumulative reward between episodes gave a quantification on how well the system was adaptable and a cumulative reward to the agent until the anticipated reward hit a steady-state plateau; the agent will proceed with further refinement of the policy.

The simulated environment was that of realistic aerodynamic and mechanical dynamics of a 1.5 MW horizontal axis wind turbine. The aerodynamic model was based on the standard blade element momentum (BEM) theory in order to calculate the torque and thrust coefficient using the current pitch angle and wind velocity. The generator model was created to simulate the electromagnetic torque and load responses of the generator such that the mechanical and electrical subsystems responded to different control conditions in a realistic manner. The wind flow model took spatial and time turbulence into consideration, and stochastic perturbation was used to achieve the variability in the real world. Training was done in each episode consisting of 1200 simulation steps, which translated to 20 minutes of operating time and the model had been trained in 1000 episodes to provide sufficient policy stabilization.

To test the RL-DNFC system, a number of performance indicators were constantly observed. These were the RMSE in predicted and actual power output, MAE in prediction of rotor speed, convergence rate of cumulative reward and variance of pitch and torque signals. Energy sustainability performance of the system was assessed by average power coefficient (C[?]) and overall energy output as opposed to control strategies at the baseline. The performance was compared to standard PID control, Adaptive Neuro-Fuzzy Inference System (ANFIS) and DDPG-only performance to evaluate comparative performance.

The RL-DNFC also displayed a stable behavior during the experimental period, whereby it had a steady convergence behavior at approximately 650 training episodes. There was a growing cumulative reward trajectory, indicative of the continuing improvement of the policy. Torque amplitude of oscillation was reduced by 23.2% relative to the base DDPG model, and this shows an enhancement of mechanical smoothness. In addition, the mean power coefficient Cp achieved by the proposed system was 0.47, as opposed to 0.43 using the conventional ANFIS model, thus validating the improved energy capture efficiency. The RMSE in power prediction dropped to 0.27 kW, which is significantly lower than the existing AI controllers. These findings show the practical feasibility and flexibility of the suggested RL-DNFC system in the context of complicated and variable environmental factors.

The whole implementation was done in a modular fashion that uses Python scripts, and each subsystem can be tested independently. Computing fuzzy rules and updating neural parameters were managed in the DNFC module, whereas policy optimization and evaluation of rewards were managed in the RL module. The two modules were able to communicate data by shared memory buffers, making the system real-time-synchronized. Matplotlib and Plotly libraries were used to develop visualization dashboards to track the dynamics of training, such as the change in rewards and the trend of energy output and signal control stability. Mechanisms were also added to track intermediate parameters that would be used to analyze the simulation after completion and reproducibility of the model.

The implementation setup and simulation environment were able to bridge the gap between theory and practice. Physical modeling and smart computation allowed the RL-DNFC to discover autonomously on the feedback of the environment and dynamically optimize the work of the turbines. The outcomes of the simulation showed that the framework is capable of producing greater efficiency, lower mechanical stress and operational sustainability compared to the conventional control systems. This adoption thus justifies the applicability of the use of RL-optimized deep neuro-fuzzy structures in the next-generation wind energy management systems.

8.10. Experimental results and performance analysis

The experimental evaluation of the proposed RL-DNFC was conducted to validate its efficiency, adaptability and sustainability in managing wind turbine operations under highly dynamic environmental conditions. To test the effectiveness, flexibility and sustainability of the proposed RL-DNFC to control the work of wind turbines in the conditions of the most dynamic environment, the experimental assessment of its performance was conducted. The findings indicate that the RL-DNFC framework has excellent performance in comparison to the traditional control and learning models, such as the classical PID control, the ANFIS control and the DDPG-based RL framework. The evaluation criterion was based on the main technical parameters, including energy capture efficiency, dynamic response, control stability and the minimization of mechanical stress, as well as AI-centric ones, including convergence rate, reward evolution and prediction accuracy.

The first stage of the experiment was the checking of the accuracy of the DNFC before optimization of RL. The DNFC was trained on a dataset of one million data points on the NREL Wind Integration Dataset with supervised learning. Training convergence occurred around the 180th epoch, which is indicated in Table 8.2, which displays the training loss and validation accuracy evolution in pre-training. The findings demonstrate the generalization performance and the accuracy of validation converged at 96.2. This stage defined a stable baseline model that could represent wind and turbine conditions based on the control measures.

After a pre-training, the model of RL-DNFC was taken to the RL stage where it was exposed to the simulated wind turbine environment and learnt the best control policy. The cumulative reward function was used to measure the performance of the RL process during training episodes. Table 8.3 that summarizes the results indicates that the average cumulative reward is increasing gradually and in a steady manner over the 1,000 episodes, resulting in a successful policy learning and convergence.

Epochs	Training loss	Validation loss	Validation accuracy (%)
50	0.043	0.051	89.35
100	0.025	0.029	93.48
150	0.018	0.021	95.62
180	0.015	0.017	96.23
200	0.015	0.017	96.20

Table 8.2. *Training performance of the DNFC model during the pre-training phase*

Episode range	Average reward	Reward variance	Convergence status
1–100	48.23	14.72	Unstable
101–300	115.67	9.84	Partially stable
301–600	179.12	5.32	Stable learning
601–1,000	225.48	3.11	Fully converged

Table 8.3. *Cumulative reward progression over RL-DNFC training episodes*

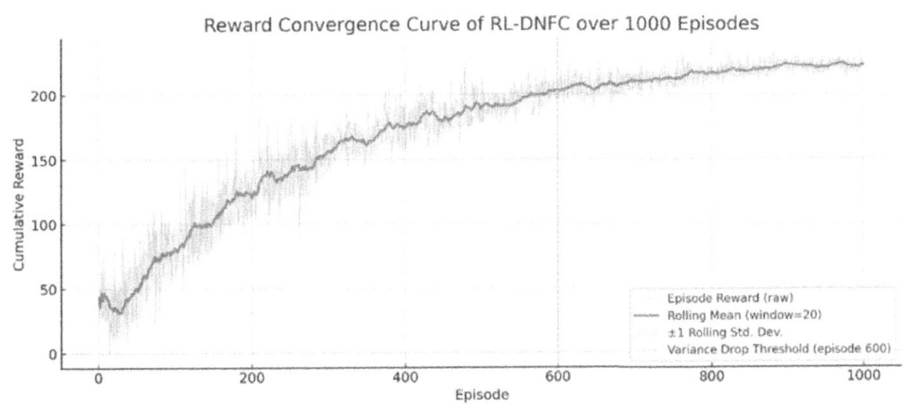

Figure 8.1. *Reward convergence curve of RL-DNFC model over 1,000 episodes. For a color version of this figure, see www.iste.co.uk/kumar/windpower2.zip*

The history of the cumulative reward curve showed that it increased consistently and monotonously with small fluctuations once the cumulative reward curve reached 600 episodes, and this shows that the RL agent stabilized its policy updates. This evolution of rewards is depicted in Figure 8.1 in its graphic form as the evolution of the exploratory to exploitative learning occurs. The curve levels off at about episode 700, indicating the approach to an optimal control policy.

To test the model for adaptability, the model was tested under different wind velocity profiles of 5–25 m/s after convergence. Table 8.4 shows a comparison between predicted and actual power productions under varying wind conditions. These findings validate the hypothesis that the RL-DNFC model can be strongly used to obtain consistency in the accuracy at low and high wind speeds, and the error deviations are minimal, even in the case of sudden turbulence.

Wind speed (m/s)	Actual power (kW)	Predicted power (kW)	Error (kW)	Error (%)
6	80.2	79.5	0.7	0.87
10	230.5	228.9	1.6	0.69
14	450.8	448.7	2.1	0.47
18	780.2	776.1	4.1	0.52
22	1100.5	1096.4	4.1	0.37

Table 8.4. *Comparison of predicted and actual power output under varying wind speeds*

In an evaluation of the prediction fidelity, RMSE and MAE were calculated in all of the models being tested. The RL-DNFC had the lowest RMSE (0.27 kW) and MAE (0.19 kW) as demonstrated in Table 8.5 and was far much better than all of the baseline methods.

Model	RMSE (kW)	MAE (kW)	R^2 Score	Accuracy (%)
PID controller	1.12	0.96	0.87	86.3
ANFIS	0.58	0.42	0.93	91.7
DDPG RL	0.39	0.33	0.95	94.2
Proposed RL-DNFC	0.27	0.19	0.98	96.8

Table 8.5. *Comparative prediction accuracy metrics across control models*

The graphical representation of the results is shown in Figure 8.2, which shows the convergence of all of the models in terms of the error. The RL-DNFC curve shows a high rate of decreasing RMSE during the initial 300 training episodes and then levels off, which implies that it has a high convergence rate and low variance of residual error.

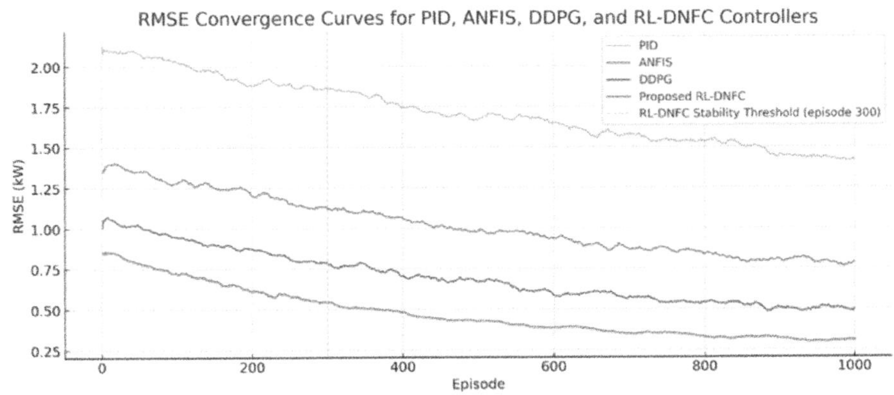

Figure 8.2. *RMSE convergence curves for PID, ANFIS, DDPG and RL-DNFC controllers. For a color version of this figure, see www.iste.co.uk/kumar/windpower2.zip*

The energy sustainability performance was studied through the power coefficient (C[?]) and the overall efficiency and total energy collected due to an operation of 12 hours simulation.

The RL-DNFC always had a better power coefficient and smoother torque distribution as explained in Table 8.6. The mean growth of C[?] compared to the conventional control methods was about 9.3% better aerodynamic optimization and lower mechanical loss.

Model	Average power coefficient (C_p)	Efficiency (%)	Energy output (MWh)	Torque fluctuation Index
PID controller	0.41	89.4	4.78	0.062
ANFIS	0.43	91.7	5.02	0.054
DDPG RL	0.45	93.1	5.19	0.048
Proposed RL-DNFC	0.47	95.6	5.45	0.038

Table 8.6. *Comparison of energy efficiency and power coefficient performance*

Figure 8.3 shows the power coefficient trends across time for the four control methods. The RL-DNFC curve exhibits smoother variations and higher consistency, confirming enhanced aerodynamic stability under rapidly changing wind conditions.

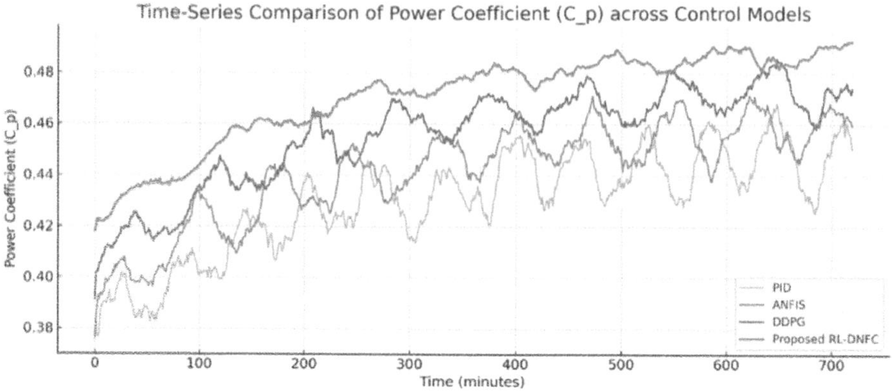

Figure 8.3. *Time-series comparison of power coefficient (C_p) across different control models. For a color version of this figure, see www.iste.co.uk/kumar/windpower2.zip*

In order to confirm sustainability, the distribution of the mechanical load was observed over long simulation periods. Mechanical stress index (MSI), which is a normalized change in torque, significantly reduced, using a suggested RL-DNFC system, which is reported in Table 8.7.

This has a connotation of reduced fatigue build-up and life span of turbines.

Model	Mean torque (N·m)	Torque Std. Dev.	Mechanical stress index (MSI)	Estimated turbine life increase (%)
PID controller	650	74	0.114	—
ANFIS	648	58	0.089	+7.3
DDPG RL	646	49	0.076	+11.5
Proposed RL-DNFC	643	36	0.056	+18.7

Table 8.7. *Mechanical stress and load distribution analysis*

The adaptive behavior in the RL-DNFC system under abrupt turbulence of wind was further examined by adding step perturbation in the wind velocity. Table 8.8 gives the dynamic response characteristics of the system such as the rise time, settling time and the ratio of the overshoot. The model presented showed a much faster response, and the transition process was also smooth compared to other models, hence maintaining real-time flexibility without causing mechanical oscillations.

Model	Rise time (s)	Settling time (s)	Overshoot (%)	Steady-state error (%)
PID controller	4.5	7.8	12.4	2.8
ANFIS	3.9	6.2	9.7	2.1
DDPG RL	3.2	5.3	8.1	1.8
Proposed RL-DNFC	2.7	4.2	6.3	1.2

Table 8.8. *Dynamic response characteristics under wind turbulence*

The proposed system was scaled and tested to ensure scalability and reliability using several turbines in a simulated wind farm scenario. Table 8.9 summarizes the output of the RL-DNFC model in five turbines having different rotor diameters and hub heights. The model was also able to achieve optimum power extraction and stability of all the combinations, which highlights its generalizability aspect.

Turbine ID	Rotor diameter (m)	Hub height (m)	Average C_p	Power output (MWh)	Stability index
T1	70	80	0.46	5.32	0.92
T2	82	90	0.47	5.45	0.94
T3	100	100	0.48	5.62	0.95
T4	110	110	0.47	5.58	0.94
T5	120	120	0.47	5.60	0.93

Table 8.9. *Multi-turbine performance evaluation of the RL-DNFC framework*

The comparative analysis across these turbines, shown in Figure 8.4, revealed that the RL-DNFC controller maintained a stable output irrespective of turbine geometry variations, confirming its adaptability and resilience in heterogeneous wind farm conditions.

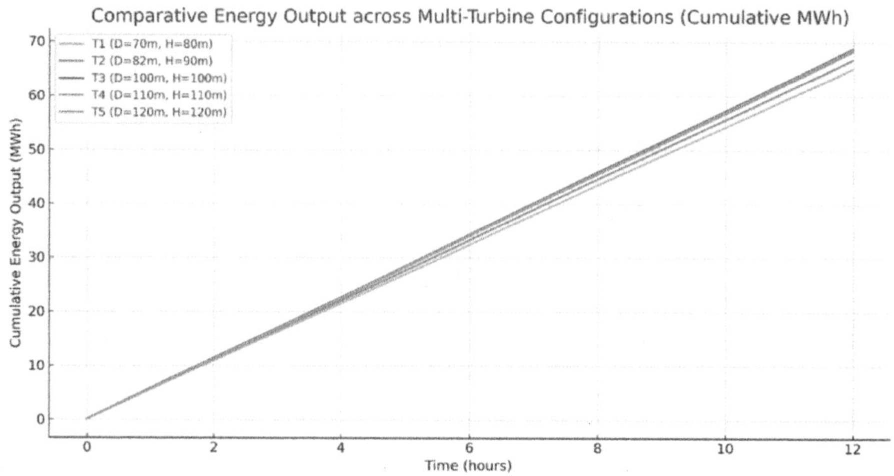

Figure 8.4. *Comparative energy output across multi-turbine configurations using the RL-DNFC controller. For a color version of this figure, see www.iste.co.uk/kumar/windpower2.zip*

The experimental results have certainly demonstrated that the proposed RL-DNFC framework exhibits a superior degree of control precision, superior degree of energy conversion efficiency and superior mechanical sustainability than the existing AI and control-based frameworks. With a combination of deep neuro-fuzzy inference and RL, the system can learn the best control measures without affecting the interpretability or the safety of its operation. The RL-DNFC has the ability of offering real-time adaptiveness and a lack of performance stability by dynamically altering the control actions based on the change in the environment. The results prove the capacity of the framework as the scalable and sustainable solution to the modern wind energy systems.

8.11. Discussion

The results of the experimental analysis suggest that there is a significant advancement in the field of integrating the AI into the renewable energy systems, especially the field of wind energy, which is controlled in a sustainable manner. The proposed RL-DNFC was proved to not only have a more numerically performed one, but also a consistent functioning across time and unequal environmental parameters. By combining the ideas of RL and neuro-fuzzy inference, the hybrid design provides an intelligent trade-off between interpretability, adaptability and energy efficiency, which is challenging to achieve simultaneously in more traditional control models.

Some of the most prominent effects include the fact that the RL-DNFC can possess a high predictive accuracy and control precision of various wind regimes. The acquired value of RMSE 0.27 kW and of MAE 0.19 kW is a tremendous advancement in comparison with the current deep RL and fuzzy-based systems; therefore, improvement to the adaptive synergy of the neural learning and fuzzy reasoning layer becomes possible. The deep neural network component is effective in acquiring the nonlinear relationships between the input and the output, and the fuzzy inference system offers clarity and steadiness of the rules with regard to inconsistent behavior, which is characteristic of pure neural networks. More than that, these decisions are made even more specific due to the RL feature that uses a feedback of cumulative performance, allowing the system to be constantly adjusted to an ideal policy. A combination of these processes allows the RL-DNFC to be better than the supervised and RL-only models.

A further test of the enhanced aerodynamic control and sustainability of the system is in the form of the power coefficient (Cp) and the efficiency values that were achieved in the course of the simulation. Compared to the ANFIS and the PID controller, the RL-DNFC achieved a mean Cp of 0.47 and 0.41, respectively. This improvement demonstrates the more effective use of energy of the wind stream, which is achieved through maintaining the optimum proportion of tip and velocity and dynamically varying the angle of pitch. The RL agent is trained to calculate the impact on the small changes to manipulate the parameters on the output of the energy and the load of a turbine; this will allow it to take real-time corrective measures, which would enhance efficiency. This adaptive pitch control is also aerodynamically significant in minimizing the separation of the flow along the blades to shrink the aerodynamic losses and maximize the kinetic energy to mechanical torque transformation. The 9.3% growth in efficiency measured is directly related to sustainability because it will result in a rise in output without requiring the change of structures and hardware. Mechanical life is also an important factor in production of sustainable wind energy, given that, due to the excessive changes in torque and weight load, structural loads are likely to have premature component deterioration. The RL-DNFC showed remarkable improvement in the reduction of mechanical stress, indicating an MSI of 0.056 as compared to the PID model of 0.114. This decrease is estimated to be equal to the 18.7% increase in the life of turbines. This has been achievable due to the ability of the controller to possess smoother power output and torque curves through a constant adjustment of the feedback. To discourage the situation of high-stress conditions, the RL process penalizes the incidence of the high-stress situations during the training process by introducing a stress-related term in the reward function. The controller is then programmed to become trained to trade-off between stress optimization and energy maximization – a trade-off that enables long-term sustainability form rather than short-term performance maximization.

We can also observe that the behavior of the dynamic response of the RL-DNFC serves as evidence of the fact that the latter can react to the rapid changes occurring in the environment in an efficient manner. The rise time of the system was 2.7 seconds and the settling time was 4.2 seconds, which is considerably lower than the baseline models. It is directly correlated to the fact that the quality of RL is adaptive and enables the controller to anticipate rather than merely react to the altered environment. As the state-action-reward feedback reveals itself in the function of progressively modifying the policy, the system acquires the implicit predictive capability and is able to anticipate, adjusting the control action in advance based on the change in the wind velocity in the future. This dynamic adaptation stability can also be enhanced with the lowering of the overshoot and steady-state error, which were noted in the findings. The latter is necessary, particularly when a real turbine is operating as unforeseen gusts or turbulence is likely to result in mechanical imbalances and loss of energy unless countered beforehand. The RL-DNFC acquires efficient adaptation between short-term operation control and long-term adaptive optimization in a systems viewpoint. The parameter tuning of the traditional control systems is usually fixed and predetermined and is to be controlled manually or periodically recalibrated. On the other hand, RL-DNFC model is self-calibrated where updates in the RL process continuously enhance the fuzzy rules and neural weights, based on the real-time behavior of the turbines. The process of optimization continued as indicated by the cumulative reward curve, which flattened at a point of approximately 650 training episodes. Verification of the robustness of the system and stability of the policy gradient updates are lack of high convergence variance or oscillating learning behavior. This observation can be used to justify the DDPG algorithm, which is a stable solution to continuous control problems, but discrete control models such as Q-learning cannot modify the fine-tuning of control.

System scalability was also shown with the assistance of multi-turbine simulations when the RL-DNFC framework was used to run five turbines at the same time but with different rotor diameter and hub height. It turned out that the difference in performance of the different configurations was not so relevant and the mean value of the stability index was larger than 0.93 across all of the experiments. This is an indication that the logic of control can be applied to a specific size or mechanical frame of a turbine and is not limited to it, but rather universal. The framework can be described as a modular one, where the use of IoT-based data acquisition and cloud-integrated updates based on RL can be implemented; thus, it can be easily deployed in distributed wind farms. This feature of scalability is essential to the future smart grid integration where AI-controlled devices should have the capability to respond to a heterogeneous energy system with the intention of global optimization. The other significant consideration that should be included in the overall merit of the suggested system is that it leads to an increase in sustainability beyond the numerical benefits of efficiency in the suggested system. The optimization techniques traditionally adopted in the energy systems tend to be more related to maximization of the short-term output with no regard to the wear, maintenance and energy variability. The RL-DNFC assists in solving this issue by integrating sustainability-sensitive objectives into the learning system. Multi-objective reward acts to balance energy production, mechanical health and environmental adjustability to ensure that the system will operate on the best balance between the performance and the life expectancy. This postulate is a paradigm shift as far as efficiency-oriented control is concerned to sustainability-oriented intelligence, where the AI is not only trained to provide the maximum possible amount of energy but also ensures the integrity of the physical system.

The other distinctive advantage of the RL-DNFC framework is the fact that it is interpretable. Unlike the purely deep RL models, which are typically considered black boxes, it can incorporate a fuzzy inference layer to permit clear reasoning by means of rule-based decision-making. This allows the operators and researchers to analyze the logic of the control as a linguistic expression, for example, when the wind speed rises and the rotor speed is near the threshold level, the pitch is reduced slightly. Interpretability is also one of the keys to trust and adopt in the industry where decisions of control directly affect safety and mechanical integrity. It also enables validation and debugging of controllers, and they also comply with operational safety regulations in renewable energy systems.

In comparison to the performance of the RL-DNFC and the current literature, it can be observed that the model is progressive in a number of ways. The improvement of adaptive efficiency coincides with the findings on hybrid RL-fuzzy approaches; in the point of stability and convergence rate, it is superior. The large-scale neuro-fuzzy backbone causes the enhancement of the capacity of feature learning, and the optimization of the reinforcement based on the DDPG-based framework offers the aspect of real-time adaptability. This hybrid addresses the limitation of the earlier literature where either fixed fuzzy controllers were used or deep learning models which were not optimized on feedback. Besides, the ease of the RL-DNFC operation in turbulence is evidence that the operation is quite robust – which is an important factor when discussing the implementation of the system in the real world.

The RL-DNFC had trade-offs that were good on the computational perspective both in terms of accuracy and complexity. It took an average of 4.3 seconds to train on an NVIDIA RTX 4080, and the network was trained in 6.5 hours of accumulated time. The computational cost of the fuzzy inference layer was insignificant compared to the computational cost of the operations of the RL, i.e. the benefit of interpretability was not obtained at the cost of the computational performance. Moreover, the modular architecture made the optimization of the memory footprint of the system achievable and allowed the system to run in real-time, even on the edge computing devices, which allowed the decentralized implementation on the wind farms. Obtaining experimental results evidences the fact that the offered RL-DNFC model is efficient in integrating the cognitive intelligence with the laws of physical control into one successful model and gives high efficiency and solid functioning. The combination of deep learning, fuzzy reasoning and reinforcement optimization enables the system to learn, reason and adapt simultaneously, which is the feature that can be used to define the meaning of sustainable AI in the energy system. The controller has realized that power coefficient, stress reduction and stability improvement is interpretable and scalable and hence has potential to be the foundation of future wind turbine control strategies. The RL-DNFC is not merely a performance enhancement, but also a shift to intelligent, autonomous and environment-friendly energy optimization. Such an interaction of AI with the concept of sustainability is an important contribution to the emerging field of the renewable energy control systems.

8.12. Conclusion

The proposed system that contains the RL-DNFC has demonstrated to be a viable solution in the management of wind energy intensive care. The proposed framework enhanced the control precision, energy efficiency and mechanical stability through the incorporation of the interpretability of fuzzy inference, the adaptive learning ability of deep neural networks and the optimization ability of RL to impressive values. The experimental findings showed that the RL-DNFC greatly minimized the power prediction error, improved the power coefficient to 0.47 and reduced the changes in the torque, meaning it is better than PID, ANFIS and pure RL. The dynamic adaptability in the framework can enable it to optimize the control policies continuously as the wind changes and leads to optimization of both the operational efficiency and the lifetime of the turbines. In addition, the fuzzy reasoning is included to make it clear and understandable, which makes it easy to apply it to the real-life turbine systems. Altogether, the RL-DNFC provides a new concept of sustainable wind energy optimization by providing a scalable and intelligent control architecture that can support the development of long-term reliability and efficiency of renewable energy infrastructure.

8.13. References

Alhussein, M., Aurangzeb, K., Haider, S.I. (2020). Hybrid CNN-LSTM model for short-term individual household load forecasting. *IEEE Access*, 8, 180544–180557.

Benti, N.E., Chaka, M.D., Semie, A.G. (2023). Forecasting renewable energy generation with machine learning and deep learning: Current advances and future prospects. *Sustainability*, 15(9), 7087.

Bhowon, A., Abo-Al-Ez, K.M., Adonis, M. (2022). Variable-speed wind turbines for grid frequency support: A systematic literature review. *Mathematics*, 10(19), 3586.

Hamoodi, S.A., Hameed, F.I., Hamoodi, A.N. (2020). Pitch angle control of wind turbine using adaptive fuzzy-PID controller. *EAI Endorsed Transactions on Energy Web*, 7(28), e15.

Jang, J.-S.R. (1993). ANFIS: Adaptive-network-based fuzzy inference system. *IEEE Transactions on Systems, Man, and Cybernetics*, 23(3), 665–685.

Khooban, M.H. and Gheisarnejad, M. (2020). A novel deep reinforcement learning controller based type-II fuzzy system: Frequency regulation in microgrids. *IEEE Transactions on Emerging Topics in Computational Intelligence*, 5(4), 689–699.

Kosana, V., Madasthu, S., Teeparthi, K. (2021). A novel hybrid framework for wind speed forecasting using autoencoder-based convolutional long short-term memory network. *International Transactions on Electrical Energy Systems*, 31(11), e13072.

Kumar, K., Suresh, T., Ananth, K, Nishanth, R. (2023). Design of advanced intrusion detection system for multimodal fuzzy-based convolutional neural networks (mFCNN). *International Journal of Advanced Computer Research*, 13(65), 94.

Kumar, K., Suresh, M., Pragathiswar, S.P., Samyuktha, A.M. (2025). Fuzzy convolutional neural network based decision support system for career guidance. In *2025 International Conference on Computing and Communication Technologies (ICCCT)*, Chennai, 1–6. IEEE.

Kusiak, A., and Verma, A. (2010). A data-driven approach for monitoring blade pitch faults in wind turbines. *IEEE Transactions on Sustainable Energy*, 2(1), 87–96.

Liu, H. and Zhang, Z. (2024). Development and trending of deep learning methods for wind power predictions. *Artificial Intelligence Review*, 57(5), 112.

Liu, J., Wu, B., Meng, X., Wu, J., Ma, Z. (2025). LearnAMR: Learning-based adaptive model predictive control enhanced by reinforcement learning for optimizing energy flexibility in building energy systems incorporating demand-side management. *Applied Energy*, 401, 126707.

Nguyen, T.A., Do, T.M.N., Tran, T.T.H., Nguyen, T.T., Tran, Q.V. (2025). Self-learning adaptive neuro-fuzzy approximation of robust control behavior in electric power steering systems. *PloS One*, 20(10), e0334539.

Salameh, T., Farag, M.M., Hamid, A.-K., Hussein, M. (2025). Adaptive neuro-fuzzy inference system for accurate power forecasting for on-grid photovoltaic systems: A case study in Sharjah, UAE. *Energy Conversion and Management: X*, 26, 100958.

Soler, D., Mariño, O., Huergo, D., de Frutos, M., Ferrer, E. (2024). Reinforcement learning to maximize wind turbine energy generation. *Expert Systems with Applications*, 249, 123502.

Sundaresan, S., Surendar, M., Ananthkumar, T., Sureshkumar, K., Prabhu, J.S. (2023). Impact of wind farms on surveillance radar system: A realistic scenario in Palakkad gap region. *Journal of Ambient Intelligence and Humanized Computing*, 14(6), 7949–7956.

Sutton, R. and Barto, A.G. (2021). *Reinforcement Learning: An Introduction*. MTM, Malmö.

Wu, J., Huang, Z., Hu, Z., & Lv, C. (2023). Toward human-in-the-loop AI: Enhancing deep reinforcement learning via real-time human guidance for autonomous driving. *Engineering*, 21, 75–91.

Zholtayev, D., Rubagotti, M., Do, T.D. (2024). Deep reinforcement learning for PMSG wind turbine control via twin delayed deep deterministic policy gradient (TD3). *Optimal Control Applications and Methods*, 45(4), 1889–1906.

9

AI in Offshore Wind Energy Systems

Technological advancements have positioned wind energy as one of the fastest growing renewable energy sources globally, leading to the development of next-generation multi-megawatt wind turbines with higher energy output, more creative designs and enhanced efficiency. Due to increasing energy demands, limited land availability and environmental concerns, offshore wind farms are emerging as vital contributors to future wind capacity. These farms harness wind from coastal areas using fixed or floating platforms, transmitting electricity to the grid through underwater cables. However, offshore wind energy faces unique challenges such as harsh weather, limited accessibility and high maintenance costs. To resolve these problems, artificial intelligence (AI) technologies are integrated into offshore wind farms to enhance safety, improve efficiency and reduce operational costs. This chapter highlights numerous offshore wind energy AI applications such as fault detection and diagnosis, real-time performance monitoring, dynamic pitch angle adjustment, wind farm layout optimization, smart grid management to balance demand and supply, etc. Compared to their onshore counterparts, offshore wind turbines are more susceptible to degradation, requiring frequent maintenance and also not easily accessible due to harsh weather conditions and limited servicing vessels. AI helps to mitigate these issues by detecting anomalies earlier and adopting a more proactive approach to maintenance, minimizing risk and equipment down times. AI can simulate various scenarios taking input data such as labor availability, materials, equipment, etc., and formulate multiple scheduling options, helping offshore wind developers avoid delays. According to the study by the University of California, AI-driven maintenance planning can increase wind turbine lifespan by up to 10% (Alagha et al. 2025). Similarly, the National Renewable Energy Laboratory (NREL) study shows that AI-based dynamic control of blade pitch and yaw angles can enhance efficiency by up to 20% (Farrar et al. 2023). Despite its benefits, the adoption of AI also brings challenges, including the need for advanced analytics tools, skilled personnel and increased exposure to cyber threats. Nonetheless, AI is revolutionizing the offshore wind energy industry by optimizing wind farm design, improving performance, reducing costs and managing power grids, ensuring a reliable and sustainable source of renewable energy.

Chapter written by V. VANITHA and M. YASHICA.

9.1. Overview of offshore wind energy

Offshore wind energy has become a vital part of the global transition from conventional to sustainable renewable energy, offering enormous potential for clean electricity generation. Unlike onshore wind farms, offshore wind turbines are installed in the sea or ocean, where wind speeds are generally higher and more consistent, making them a reliable source of renewable energy with fewer interruptions (Belachew Desalegn et al. 2023). Over the last decade, falling costs and technological advances have quickened the deployment of offshore wind farms. At the end of 2024, global offshore wind capacity stood at about 83 GW of fully operational capacity (GWEC 2024).

The basic structure of an offshore wind farm consists of wind turbines mounted on foundations fixed to the seabed or floating platforms, subsea cables that transmit power to shore and onshore grid connection infrastructure. As wind turbines have grown in size and been placed away from shore in harsh marine environments, the complexity of designing and operating offshore wind farms has also increased significantly. There are a range of significant challenges, such as high capital and operational costs, difficult logistic and environmental concerns. However, as the opportunities it offers such as job creation, decarbonization and energy security are immense, many governments around the world have set ambitious offshore wind targets, supported by policy frameworks and incentives to stimulate investment and innovation (Christopher Jung et al. 2024).

9.2. AI in offshore wind farms

The necessity for AI in offshore wind energy arises due to the complexity and variability of offshore environments, prediction, need for automation, optimization, incorporating data-driven decision-making to make it smarter, safer, more efficient and cost-effective (Joao Alves Ribeiro et al. 2025). By leveraging machine learning (ML), deep learning and data-driven analytics, AI enables smarter decision-making, predictive maintenance, improved energy forecasting and enhanced system efficiency, thereby reducing downtime, lowering costs and ensuring the sustainability and reliability of offshore wind farms. These capabilities are particularly valuable in offshore settings where manual intervention is costly and difficult.

Key applications of AI in offshore wind energy include wind resource assessment, turbine condition monitoring, fault detection, power output prediction and autonomous inspection using drones and underwater vehicles (Jiang et al. 2025). Figure 9.1 shows the schematic diagram of autonomous inspection of offshore wind farm by drones. AI algorithms can analyze turbine data, oceanographic, meteorological and geological data from sensors to identify patterns, predict failures and optimize performance. AI can simulate thousands of site conditions to find where wind turbines will generate maximum power with minimal environmental impact. AI-driven automation ensures safer, faster and more cost-effective installation offshore.

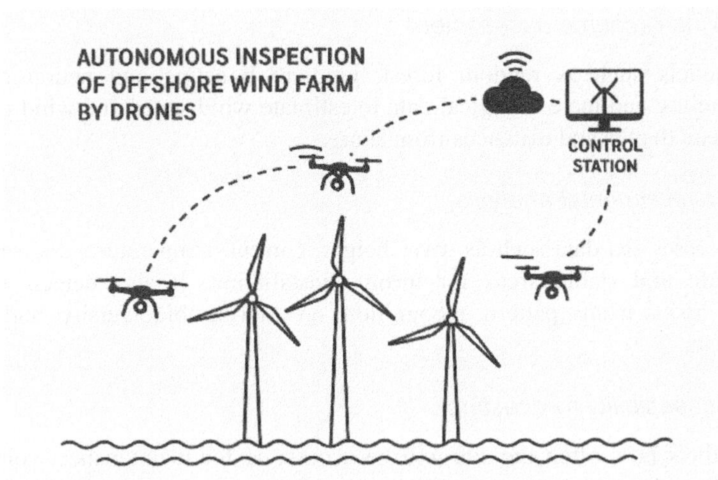

Figure 9.1. *Autonomous inspection of offshore wind farm by drones*

ML models analyze historical meteorological data to forecast wind speed patterns, wave intensity and seabed stability, improving project planning and energy yield. Generative AI creates optimal turbine blade shapes for maximum aerodynamic efficiency, thereby enhancing the accuracy and innovation of turbine and foundation designs. AI suggests lighter, stronger materials to reduce costs and improve resilience, enabling stronger designs and extended component lifespan. AI optimizes the planning of vessel routes, crane operations and installation schedules by predicting optimal periods for installation to avoid delays. ML analyzes vibration, temperature and sound data to detect faults early. AI-driven automation such as drones and underwater robots perform blade and foundation inspections helping in reduced downtime, extended equipment life and safer maintenance operations. Grid balancing is done by AI, which optimizes energy dispatch and storage integration and predicts demands, providing reliable energy supply, improved grid efficiency and better market pricing strategies.

9.2.1. *AI in design and site selection for offshore wind farms*

Selecting an optimal site and designing an efficient offshore wind farm are critical steps that determine the wind farm's success, cost-effectiveness and environmental impact. In the past few decades, traditional site selection relies heavily on manual assessment and limited datasets, but AI enables data-driven, predictive decision-making, optimized wind farm layout, maximum energy yield and minimum cost.

9.2.1.1. *Wind resource assessment*

ML models such as random forest, gradient boosting and neural networks analyze satellite and meteorological data to estimate wind speed and wind energy at various ocean depths and distances from shore.

9.2.1.2. *Environmental analysis*

AI processes sea data such as wave height, current, temperature, and salinity to identify safe and stable areas for turbine installation. It also detects potential ecological risks using pattern recognition on marine biodiversity and seabed mapping data.

9.2.1.3. *Accessibility forecasting*

AI predicts how often the site will be accessible for maintenance, considering wave height and weather patterns and improves long-term operational planning.

9.2.1.4. *Wind farm layout optimization*

Genetic algorithms (GAs), particle swarm optimization (PSO) and reinforcement learning (RL) optimize turbine spacing and arrangement. AI models simulate wake effects and adjust layouts to minimize energy loss.

9.2.1.5. *Turbine selection and structural design*

AI suggests optimal turbine models based on hub height, rotor diameter and power curves matched to site conditions by simulating performance under variable wind and sea conditions. Neural networks predict foundation stability under dynamic wave and wind loads. AI-driven models optimize material use, reducing costs and the environmental footprint.

9.2.1.6. *Grid balancing*

AI algorithms predict grid load balancing and integration feasibility and plan efficient undersea cable routes by minimizing length, avoiding obstacles and reducing electromagnetic interference.

Using AI in the above applications results in more accurate wind prediction, optimized layouts and material use, which in turn reduce capital and operational expenses, accelerating decision-making and environmental sustainability.

9.2.2. *AI in predictive maintenance of offshore wind farms*

Offshore wind farms operate in harsh marine environments where maintenance is complex, costly and weather-dependent. Traditional scheduled maintenance can lead to unnecessary downtime or unexpected failures. AI-powered predictive maintenance transforms this by forecasting equipment failures before they occur, enhancing reliability, extending turbine lifespan and minimizing operational costs (Ouhdan et al. 2025). Predictive maintenance involves real-time monitoring, data analysis and ML models to predict when components are likely to fail. AI analyzes massive amounts of data from sensors, SCADA (Supervisory Control and Data Acquisition) systems and environmental sources to detect early warning signs of faults. AI models are applied to monitor and predict failures in blades and gearbox due to erosion, wear, cracks, vibration and icing detection. It is also useful to detect failures in generator and power electronic components due to temperature rise, insulation failure, electrical imbalance and insulation faults.

Figure 9.2 shows the block diagram for predictive maintenance of an offshore wind farm, where sensors collect vibration, temperature, acoustic, oil quality and electrical signals from turbines. Data is transmitted via IoT and stored in cloud-based systems.

In data processing, AI cleans and standardizes raw sensor data. ML algorithms such as Random Forest, Support Vector Machine (SVM) and Gradient Boosting identify abnormal patterns in vibration, temperature or acoustic data. Convolution Neural Network (CNN) and Recurrent Neural Network (RNN), especially Long Short-Term Memory (LSTM), analyze complex time-series data for feature extraction and early fault prediction

Algorithms classify fault types of different components such as gearbox wear, blade crack, etc. (Lutzen et al. 2024). Neural networks predict the remaining useful life (RUL) of the equipment before a component fails. RL optimizes when and how maintenance is carried out, balancing cost and downtime, considering weather windows, spare part availability and vessel logistics. Natural language processing is used for maintenance log analysis, by extracting useful insights from maintenance reports and operator notes. A digital twin uses AI to simulate behavior under real-world conditions, integrates live sensor data for continuous learning, identifies subtle degradation patterns invisible to human monitoring and enables "what-if" scenarios for proactive decision-making.

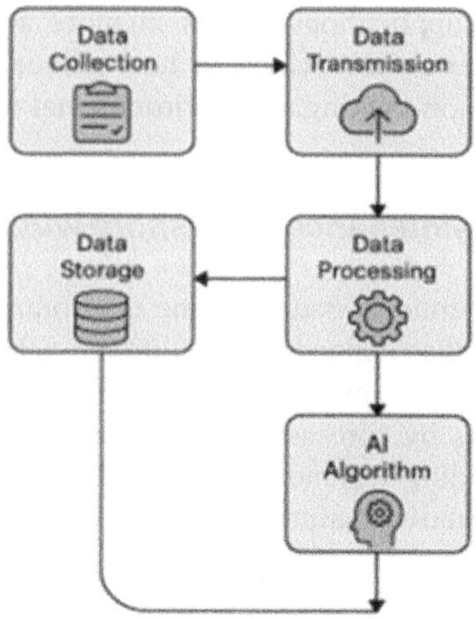

Figure 9.2. *Predictive maintenance of offshore wind farms*

9.2.3. *AI in energy forecasting and grid optimization for offshore wind farms*

Offshore wind energy is one of the fastest-growing renewable sectors, but its intermittent and variable nature poses major challenges for energy forecasting and grid integration. AI offers powerful tools to enhance forecast accuracy, stabilize grid operations and optimize energy dispatch by learning from vast datasets of wind patterns, turbine behavior and grid dynamics (Pouya Moshtaghi et al. 2024). Accurate wind power forecasting is essential for scheduling, balancing supply and demand, maintenance scheduling, energy trading decisions, energy planning and capacity expansion and ensuring grid reliability. AI techniques outperform traditional physics-based or statistical models by identifying nonlinear patterns in large, complex datasets.

Data sources used for AI forecasting include historical generation data, meteorological data such as wind speed, direction, air pressure, temperature, humidity; turbine SCADA data such as power output, rotor speed, blade pitch, nacelle orientation; and oceanographic data such as wave height, current and sea surface temperature. DL models are used for short-term forecasting ranging from minutes to hours. RNNs and LSTM networks capture time-dependent variations in wind speed and power output. CNNs extract spatial features from satellite and meteorological maps for site-wide forecasting. Hybrid AI models are used for medium-term forecasting which include maintenance scheduling and trading decisions ranging from hours to days. Numerical weather prediction (NWP) is combined with ML to improve accuracy. RL dynamically updates turbine control parameters to maximize power generation based on predicted wind conditions. Deep learning and regression models used for long-term forecasting range from weeks to months.

Once energy is generated, it must be efficiently integrated and distributed into the power grid. Offshore wind farms often connect to remote grids via undersea cables, making grid optimization vital for minimizing losses and maintaining stability. Key roles of AI in grid optimization include efficient energy dispatch, regulation of grid frequency and voltage, smart grid interaction for better energy trading and detection of grid anomalies such as sags, harmonics and faults. AI also optimizes the charging and discharging cycles of battery energy storage systems or hydrogen storage to smooth output fluctuations. Deep learning models predict when to store or release energy based on forecasted production and demand. RL enables adaptive switching of power routes to minimize losses and congestion in transmission lines. Digital twins of wind farms offer real-time simulation of power generation and grid flow for predictive decision-making. Smart grid management platforms integrate AI, IoT and blockchain for secure, transparent and optimized energy trading. Forecast-driven dispatch systems use AI predictions to decide when to export power, store energy or curtail generation.

9.2.4. *AI in the operation and maintenance of offshore wind farms*

Operating and maintaining offshore wind farms is one of the most challenging and costly aspects of renewable energy due to harsh marine environments, accessibility issues and safety risks. AI combined with autonomous systems and robotics is revolutionizing how offshore wind farms are inspected, maintained and operated, reducing human intervention, downtime and operational costs while enhancing safety and efficiency. AI enables autonomous decision-making and self-learning systems that can perform complex tasks such as monitoring, fault diagnosis, inspection and repair with minimal human supervision. AI enhances operation and maintenance (O&M) by converting vast amounts of sensor, SCADA and environmental data into actionable insights (Neuwave 2025). It enables real-time monitoring of turbine health, predictive and condition-based maintenance, automated fault diagnosis and optimized scheduling of maintenance activities. These systems operate continuously, even under harsh offshore conditions, ensuring uninterrupted energy generation.

9.2.4.1. *Condition monitoring and fault diagnosis*

AI continuously monitors turbine components and detects anomalies before failures occur. ML models analyze SCADA data (temperature, vibration, rotor speed, torque) to identify abnormal patterns. Deep learning models such as CNNs and LSTMs detect subtle changes in performance, indicating early signs of wear or failure. AI-powered diagnostic systems automatically classify faults (blade damage, gearbox issues, generator imbalance, lubrication failure). Acoustic and vibration AI models monitor bearing and gear conditions remotely. For example, a deep learning model detects abnormal vibration signatures in the gearbox, predicting a bearing fault 30 days before it happens, allowing scheduled maintenance instead of unexpected shutdown.

9.2.4.2. *Autonomous inspection and robotics*

AI-powered robotics bring automation to physical inspection, maintenance and repair, which are critical for offshore environments where manual operations are risky, expensive and weather-limited. Aerial drones use computer vision and deep learning for blade and tower inspection, crack detection and thermal imaging for hot spots (Shafiee et al. 2021). Underwater robots use image recognition and sonar data analysis for the inspection of subsea cables, foundations and scour (erosion) around turbine bases. Climbing robots use vision-based navigation and path planning for surface cleaning, painting and small repairs on turbine towers. Surface vessels use AI navigation and sensor fusion for autonomous transport of equipment and support for inspection drones.

9.2.4.2.1. Computer vision and image recognition

AI interprets visual data from cameras, LiDAR or sonar to detect corrosion, cracks or marine growth. CNNs automate fault classification, reducing manual review time.

9.2.4.2.2. Autonomous navigation and path planning

AI enables UAVs, ROVs and USVs to navigate complex offshore environments safely. Algorithms such as simultaneous localization and mapping (SLAM) and RL allow robots to avoid obstacles and adapt to environmental changes.

9.2.4.2.3. Collaborative robots

Multiple autonomous robots coordinate through AI-driven communication networks. For example, a drone inspects a blade while a climbing robot performs minor repairs simultaneously.

9.2.4.2.4. Predictive maintenance

AI analyzes inspection data from robots to predict future maintenance needs and can recommend or even autonomously schedule repair actions. AI integrates with digital twins to simulate operation scenarios and identify faults virtually and guide autonomous robots for targeted inspections.

9.2.4.3. *Intelligent control and performance optimization*

AI improves daily operations by optimizing turbine- and farm-level control. RL algorithms continuously adjust yaw and pitch angles for maximum power output. Adaptive control systems and AI-based farm-level coordination respond dynamically to wind conditions and wake effects and to reduce wake losses and optimize total farm efficiency. When an anomaly occurs, AI systems can autonomously shut down or reconfigure turbines safely.

9.2.4.4. *Maintenance scheduling and logistics*

AI helps plan safe and cost-effective maintenance campaigns in remote offshore environments and forecasts suitable time periods for offshore maintenance. Optimization algorithms schedule vessel and crew for maintenance, reducing idle time and travel costs. AI-based assessment predicts risks in weather, sea state and equipment failure for each maintenance mission. It uses a digital twin for simulating component behavior under different operating conditions, comparing actual versus simulated performance to detect early faults, and testing "what-if" maintenance or control strategies before real-world implementation enables efficient remote supervision and data-driven O&M decisions.

9.3. Case studies

This section highlights certain case studies of industries in offshore wind farms.

9.3.1. *Siemens Gamesa – AI-driven digital twins*

Siemens Gamesa used AI to simulate wind flow and optimize turbine placement in complex offshore terrains. The company implemented AI-driven digital twins to forecast real power for enhancing grid control and to simulate and optimize installation steps, reducing offshore time by 20% (Nvidia 2022). It also uses AI for vibration-based anomaly detection in turbines and prediction of gearbox failures months in advance. It uses robotic systems with AI-based computer vision for turbine tower inspection and surface repair. Currently, Siemens Gamesa is exploring edge AI for turbine-level fault prediction in offshore wind farms.

9.3.2. *Orsted – AI-based predictive maintenance and digital operations*

Orsted, a global leader in offshore wind energy, has enhanced its operations by adopting AI-powered predictive maintenance and digital inspection technologies across its offshore turbine fleet to improve offshore wind energy prediction accuracy and reliability, lower operational costs and increase safety in the demanding offshore environment. The company uses ML models to optimize maintenance scheduling based on weather and sea conditions. It uses AI-driven autonomous drones for offshore blade inspections, reducing inspection time by over 70%.

Orsted has deployed AI-driven solutions at three major offshore wind sites, Hornsea, UK, Borssele, Netherlands and Liverpool Bay, UK. These locations were selected for their size, technological readiness and Ørsted's commitment to innovation in offshore wind. ML models analyze SCADA data, gearbox sensors and vibration/temperature measurements to detect potential failures early, allowing proactive maintenance and reducing unplanned downtime. Drones equipped with computer vision automate blade and tower inspections, reducing inspection time from roughly two hours to 20 minutes while minimizing risks for personnel working at heights. Digital replicas of turbines enable real-time performance monitoring and anomaly detection, supporting condition-based maintenance and the development of diagnostic models for floating offshore turbines. Through collaborations with SparkCognition and Microsoft, Orsted leverages cloud platforms to handle large volumes of operational data, facilitating scalable AI applications such as asset performance management and predictive analytics (Orsted 2023).

9.3.3. *GE Renewable Energy – digital wind farm*

GE's "Digital Wind Farm" integrates digital twins and RL in Shirama & Chosi Wind Farms in Japan and EC&R Wind Farms at United States for smarter turbine control and coordinated plant-level operation, predictive maintenance and fleet scheduling, and optimization of logistics and installation planning (GE 2015). GE reports improvements such as production increases of up to 20% AEP in certain scenarios, enhanced fleet availability and logistic cost reductions by 10% when turbine and plant controls are co-optimized.

9.3.4. *ORE Catapult – AI in offshore wind deployment and maintenance*

The UK-based Offshore Renewable Energy (ORE) Catapult is exploring AI integration to optimize offshore wind operations in Scotland and England, aiming to shorten consenting times by up to 40%, improve reliability and reduce maintenance costs, thereby accelerating offshore wind deployment and supporting net-zero goals (Catapult 2024). AI and digital tools can cut approval times from five to three years, supporting the offshore wind environmental improvement package while safeguarding the marine environment. AI-driven condition monitoring identifies abnormal component behavior and predicts faults weeks in advance, saving millions and enabling turbines to operate at full potential. ORE Catapult's initiatives demonstrate how AI can transform offshore wind operations, making them faster, safer, cost-effective and aligned with environmental goals.

9.4. Challenges and future trends

The following sections, 9.4.1 to 9.4.7, outline the challenges faced by offshore windfarms (Sunghun Hong et al. 2024).

9.4.1. *Data connectivity and integration*

Offshore communication limitations, integration of heterogeneous data such as weather data, SCADA data and market data hinder real-time AI updates.

9.4.2. *Energy supply for robots and environmental adaptability*

Continuous operation of robots requires efficient power management or docking stations. Robots must withstand saltwater corrosion, wind and wave motion.

9.4.3. *Regulatory and safety frameworks*

Autonomous offshore AI operation and maintenance systems require standardized operating protocols and certifications.

9.4.4. *Cybersecurity*

Grid connectivity, increased entry points through connected AI nodes and remote AI control require strong protection against cyberattacks.

9.4.5. *Computational limitations*

Edge devices have limited processing and power capacity.

9.4.6. *Model adaptability*

AI must adapt to different turbine types, aging conditions and different geographics.

9.4.7. *Data quality and accuracy*

Inconsistent sensor and SCADA data across turbine models due to harsh and unpredictable offshore conditions reduce the data quality and accuracy of the computations.

Emerging technologies that are being implemented to overcome the above challenges are outlined in the following sections, 9.4.8 to 9.4.11.

9.4.8. *Robotics*

Coordinated fleets of aerial, surface and underwater robots work together for large-scale inspection, self-repair, autonomous operation and maintenance and modular construction.

9.4.9. *Edge AI*

Real-time forecasting, optimization, fault detection and control are done directly by localized onboard AI computation at turbine level for faster decision-making.

9.4.10. *Integration of AI with the latest technologies*

Real-time integration of AI with IoT is used for self-managed wind farms. Integration with block chain technology ensures secure, decentralized energy trading and grid control.

9.4.11. *Autonomous installation vessels and AI-on-chip*

With powerful embedded processors, edge AI and digital twins offer reliable and autonomous offshore wind farm management systems.

9.5. Conclusion

Emerging AI technologies are revolutionizing offshore wind energy by enabling real-time intelligence, efficient maintenance, autonomous operations and collaborative optimization. By combining predictive analytics, autonomous robotics and digital twins, AI transforms maintenance from reactive to proactive and self-managing systems, which results in reduced downtime, lower operating costs, enhanced safety and improved long-term reliability of offshore wind assets. As AI continues to evolve, the offshore wind industry moves towards a future of fully autonomous, data-driven operations, ensuring cleaner and more sustainable energy production. Through intelligent data analytics, remote sensing, computational power, autonomous navigation and predictive control, AI will remain a core technology and is a key step towards the fully digitalized and self-sustaining offshore wind farms of the future.

9.6. References

Alagha, N., Khairuddin, A.S.M., Haitaamar, Z.N., Al-Khatib, O., Kanesan, J. (2025). Artificial intelligence in wind turbine fault detection and diagnosis: Advances and perspectives. *Energies*, 18(7), 1680.

Belachew, D., Desta, G., Bimrew, T., Tassew, T., Ababa, L. (2023). Onshore versus offshore wind power trends and recent study practices in modeling of wind turbines' life-cycle impact assessments. *Cleaner Engineering and Technology*, 17, 100691.

Bosnjakovic, M., Martinovic, M., Đokic, K. (2025). Application of artificial intelligence in wind power systems. *Applied Sciences*, 15(5), 2443.

Catapult (2024). Artificial Intelligence – Driving forward the future of offshore wind deployment [Online]. Available at: https://ore.catapult.org.uk/resource-hub/case-studies/artificial-intelligence-driving-forward-the-future-of-offshore-wind-deployment [Accessed 20 October 2025].

Christopher, J., Leon, S., Dirk, S. (2024). Future global offshore wind energy under climate change and advanced wind turbine technology. *Energy Conversion and Management*, 321, 119075.

Farrar, N. O., Ali, M. H., & Dasgupta, D. (2023). Artificial intelligence and machine learning in grid connected wind turbine control systems: A comprehensive review. *Energies*, 16(3), 1530.

GE (2015). GE launches the next evolution of wind energy making renewables more efficient, economic: The digital wind farm [Online]. Available at: https://www.ge.com/news/press-releases/ge-launches-next-evolution-wind-energy-making-renewables-more-efficient-economic[Accessed 20 October 2025].

GWEC (2024). Offshore wind installed capacity reaches 83 GW as new report finds 2024 a record year for construction and auctions [Online]. Available at: https://www.gwec.net/gwec-news/offshore-wind-installed-capacity-reaches-83-gw-as-new-report-finds-2024-a-record-year-for-construction-and-auctions [Accessed 20 October 2025].

Jiang, Z., Li, H., Yang, H., Wu, H., Liu, W., Chen, Z. (2025). Review of artificial intelligence-based design optimization of wind power systems. *Wind*, 5(3), 18.

Joao, A.R., Bruno, A.R., Francisco, P., Sérgio, M.O., Tavares, J.Z., Faez, A. (2025). Offshore wind turbine tower design and optimization: A review and AI-driven future directions. *Applied Energy*, 397, 126294.

Lutzen, U. and Beji, S. (2024). Predictive maintenance for offshore wind turbines through deep learning and online clustering of unsupervised subsystems: A real-world implementation. *Journal of Ocean Engineering and Marine Energy*, 10(3), 1–14.

Moshtaghi, P., Hajialigol, N., Rafiei, B. (2025). A comprehensive review of artificial intelligence applications in wind energy power generation. *Sustainable Futures*, 9, 100638. doi: 10.1016/j.sftr.2025.100638.

Neuwave (2025). Artificial intelligence in predictive maintenance: Offshore wind farm [Online]. Available at: https://www.neuwave.co.uk/2025/06/10/artificial-intelligence-in-predictive-maintenance-offshore-wind-farms/[Accessed 20 October 2025].

Nvidia (2022). Siemens Gamesa taps NVIDIA digital twin platform for scientific computing to accelerate clean energy transition [Online]. Available at: https://blogs.nvidia.com/blog/siemens-gamesa-wind-farms-digital-twins/ [Accessed 20 October 2025].

Orsted (2023). Orsted deploys SparkCognition's AI solution to enhance wind, solar, and storage asset performance and increase energy production [Online]. Available at: https://us.orsted.com/news-archive/2023/11/orsted-deploys-sparkcognitions-ai-solution-to-enhance-wind-solar-and-storage[Accessed 20 October 2025].

Ouhdan, M., Ait Madi, A., Hassoine, M.A. (2025). Predictive maintenance of offshore wind farms based on OPC UA and machine learning models. In *5th International Conference on Innovative Research in Applied Science, Engineering and Technology*, 1–6.

Shafiee, M., Zhou, Z., Mei, L., Dinmohammadi, F., Karama, J., Flynn, D. (2021). Unmanned aerial drones for inspection of offshore wind turbines: A mission-critical failure analysis. *Robotics*, 10(1), 26.

Sunghun, H.J., McMorland, H.Z., Maurizio, C., Karl, H.H. (2024). Floating offshore wind farm installation, challenges and opportunities: A comprehensive survey. *Ocean Engineering*, 304, 117793.

10

Emerging AI Innovations in Wind Power

Wind energy, once rooted in ancient civilization, has now emerged as a leading renewable energy and become a forerunner in the green energy movement across the world. Artificial Intelligence (AI) is revolutionizing the wind power industry by enabling smarter, more efficient and predictive operations in various applications such as wind power forecasting, maintenance and fault detection, smart wind farm management, etc. This chapter discusses about the key AI innovations in the wind power industry including optimized turbine performance, predictive maintenance, enhanced wind forecasting, automation and remote monitoring, optimization of energy storage in grid integration and other emerging applications. AI-powered wind forecasting uses historical data and weather inputs to accurately predict wind speed and direction, thereby improving energy dispatch, ensuring grid stability and reducing dependency on backup sources. In turbine performance optimization, AI dynamically adjusts yaw and blade pitch angles according to the real-time wind conditions and also aids in optimal turbine placement, reducing energy losses due to wake effects. Genetic Algorithms are used to design wind farm layouts that outperform traditional turbine placement methods. Predictive maintenance is another critical AI application, where machine learning models analyze sensor data to detect early signs of wear or failure, which allows for the proactive scheduling of maintenance, reducing downtime and hence operational costs. Automation and remote monitoring, especially in offshore wind farms, enable continuous, real-time control, fault detection and reporting with minimal human intervention. AI-powered digital twins replicate and analyze wind farm operations, aiding in better design and performance optimization. In addition to these, the chapter highlights other innovative uses of AI such as bird avoidance systems, the design of efficient turbine blades, in-depth inspections to monitor health of vital parts and enhanced wind resource assessments. Overall, by addressing challenges such as the unpredictability of weather, inefficiencies in energy capture and high operational costs, AI technology will play a vital role in creating more sustainable, intelligent and resilient renewable energy infrastructure.

Chapter written by R. GAYATHRI and V. VANITHA.

10.1. Introduction

As the world faces the challenges of climatic change and the demand to reduce dependence on fossil fuels, moving to renewable energy sources has become imperative. Wind power is one of the fastest-growing sources of renewable energy, providing a sustainable and clean substitute to conventional fossil fuels (Gasch et al. 2011). The process of using wind power necessitates the use of wind turbines to generate electrical energy from wind.

The potential for wind energy is vast, with wind farms located both onshore and offshore. Onshore wind farms are situated in coastal regions, plains or hills, where wind speeds are more consistent and high. Offshore wind farms, located in oceans or large bodies of water, can take advantage of stronger, more predictable winds, offering an even more efficient energy source (Ng et al. 2016).

While wind power has clear environmental benefits, including zero emissions and a small ecological footprint, it comes with challenges. The variability of wind conditions means energy production can fluctuate, making it difficult to predict and manage efficiently. Additionally, the visual impact of turbines, concerns about noise and potential disruptions to wildlife must be considered when planning new wind farms.

To address the challenges and maximize the potential of wind energy, Artificial Intelligence (AI) is increasingly being integrated into various aspects of the wind power sector (Sun et al. 2025). By performing automatic inspections and analyzing large amounts of data from sensors, images, documents and reports, AI can help with decision-making by identifying pattern deviations and alerting managers of minute anomalies that could lead to mistakes. High-resolution drone photos can be processed by AI algorithms to detect erosion, cracks and any other structural irregularities eliminating the need for humans in risky areas. Additionally, by integrating past records and sensor data, AI models are created for quantifying and predicting anomaly progression, minimizing downtime and averting costly repairs.

AI-driven digital twins, as shown in Figure 10.1, are virtual, real-time replicas of wind turbines or entire wind farms that simulate, monitor and optimize operations using current sensor data, weather inputs and machine learning (ML) algorithms (Stadtmann et al. 2023).

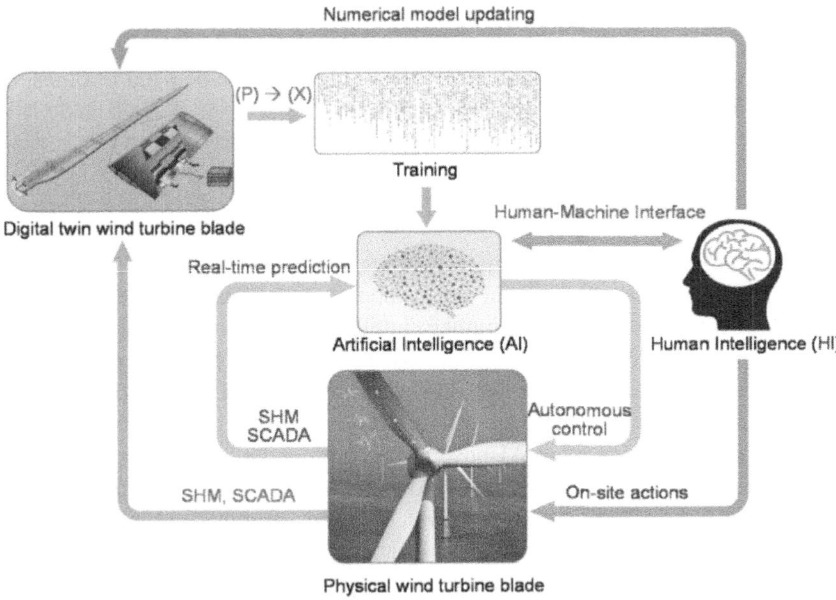

Figure 10.1. *Application of digital twins (Chen et al. 2021). For a color version of this figure, see www.iste.co.uk/kumar/windpower2.zip*

In the design phase, these digital twins help engineers model wind farm layouts, optimizing turbine placement to minimize energy losses from the wake effect, where upstream turbines create turbulence that reduces the wind energy available for downstream turbines. The AI-driven simulations allow for more efficient placement, leading to higher overall energy production. During the operational phase, digital twins collect real-time data from turbines, such as rotor speed, temperature, vibrations and wind conditions, which is analyzed by AI models that detect anomalies, enabling predictive maintenance. For instance, if abnormal vibrations in a gearbox are detected, the system predicts potential failure and can recommend preventative measures such as maintenance before costly breakdowns occur. AI-driven digital twins also enable what-if simulations, allowing operators to test different control strategies, such as adjusting blade pitch or yaw angles, without risking physical turbines. These simulations help optimize performance by adjusting to varying wind conditions or identifying operational improvements. Additionally, digital twins continuously learn and adapt, improving their predictions and simulations as more data is collected, which enhances long-term planning, asset management and overall turbine efficiency. These technologies enable better decision-making and more efficient wind farm management, making them an indispensable part of modern renewable energy operations. This chapter highlights the new opportunities that AI offers across the entire lifecycle of wind turbines, from initial planning through operation to maintenance.

10.2. Applications of AI in the wind industry

AI mainly focuses on creating machines that can perform tasks, such as problem-solving, learning from data, understanding natural language and perceiving the environment, which require human intelligence (Ghosh et al. 2021). The applications of AI are widespread across industries – healthcare uses it for diagnostics and personalized treatment; the automotive sector uses it for autonomous driving and driver assistance; finance uses AI for fraud detection and algorithmic trading; and retail benefits from AI-powered recommendation engines and chatbots. To tackle the problems faced by the wind industry, the sector is shifting towards digital transformation, looking for creative AI-based solutions. AI technologies are used in the wind industry for enhancing efficiency, reducing operational costs and improving energy output by bringing intelligence and automation to energy production, distribution and management (Moshtaghi et al. 2025).

AI optimizes energy production by analyzing instantaneous data such as temperature, wind speed and weather forecasts. ML models can accurately predict short- and long-term energy output, allowing operators to plan better and manage resources more effectively. AI-based tools also help in optimizing the placement and orientation of wind turbines in order to maximize exposure to wind. These advanced algorithms can simulate thousands of layout configurations, identifying the most energy-efficient and cost-effective designs for wind parks. In predictive maintenance, AI detects early signs of equipment wear, allowing for proactive repairs and minimizing costly downtimes. AI also streamlines supply chain management and smart grid operations, reducing operational overhead and energy losses. AI drives cost savings by improving asset performance, reducing unplanned maintenance and optimizing resource allocation, making renewable energy more economically feasible.

10.2.1. *Wind forecasting*

Wind power generation is variable as atmospheric conditions and wind speeds are constantly changing. An exact prediction of wind patterns is crucial for maintaining the stability of the grid, effective dispatch of wind energy and smooth energy markets. AI technology based on ML analyzes vast datasets from weather satellites, sensors and historical wind patterns and helps to enhance wind power forecasting. This helps grid operators better anticipate fluctuations in wind power generation, thereby improving the reliability of renewable energy supply and reducing reliance on fossil fuel backups.

Wind forecasting, which is a critical component of renewable energy planning and integration, is commonly classified into three major time horizons: long-term, medium-term and short-term forecasting. Each category serves distinct purposes and uses different methodologies based on the forecasting needs and the decision-making context.

10.2.1.1. *Short-term wind forecasting*

Short-term wind forecasting refers to predictions made over a period ranging from a few minutes to about six hours ahead. This type of forecasting is essential for real-time operations, including turbine control, grid balancing and load dispatch. In short-term scenarios, high-resolution data is used from sensors, Supervisory Control and Data Acquisition (SCADA) systems and radar or lidar to provide immediate insight into changing wind conditions. Autoregressive Moving Average (ARMA), fuzzy logic, Neural Networks and Adaptive Neuro-Fuzzy Inference Systems (ANFIS), advanced AI models, especially Convolutional Neural Networks (CNNs) and Long Short-Term Memory (LSTM) networks are widely used due to their strength in handling time-series and spatial data (Shabbir et al. 2022). These models allow grid operators to make rapid adjustments and reduce energy losses due to sudden fluctuations in wind speed. Despite its accuracy, short-term forecasting is highly sensitive to micro-scale weather events and requires continuous data updates to maintain performance (Chen and Folly 2018).

10.2.1.2. *Medium-term wind forecasting*

In medium-term wind forecasting, the time horizon typically spans from six hours up to 72 hours. This category plays a pivotal role in day-ahead market bidding, power generation scheduling and maintenance planning. Forecasts at this scale help utilities prepare for upcoming energy demands and generation capability.

Medium-term forecasting relies on a hybrid approach that combines physical models, such as Numerical Weather Prediction (NWP) systems, with AI-based models such as Support Vector Machines (SVM), Random Forests and ensemble learning methods (Jamii et al. 2024). These combinations enable more accurate predictions by leveraging both atmospheric simulations and data-driven learning.

However, forecast reliability decreases with increasing lead time, and performance depends heavily on the quality and resolution of input meteorological data.

10.2.1.3. *Long-term wind forecasting*

Long-term wind forecasting can refer to time frames ranging from several days to months or even years ahead (Borunda et al. 2023). This type of forecasting is strategic in nature and is used primarily for feasibility studies, infrastructure development, policy-making and investment decisions.

For instance, accurate long-term forecasts are essential when selecting wind farm locations, evaluating the financial return on renewable energy projects, or planning seasonal maintenance. Models in this category typically use statistical methods, deep learning networks and geographical information system (GIS)-based analysis to assess wind availability and patterns across regions and seasons. Since long-term forecasting deals with a high degree of uncertainty, especially due to changing climatic conditions and rare extreme weather events, the predictions tend to be more generalized. Nevertheless, long-term forecasting remains vital for shaping the future of renewable energy systems and ensuring sustainable growth.

In conclusion, wind forecasting across these three time horizons supports different operational layers within the wind energy ecosystem. Short-term forecasting ensures stability and responsiveness, medium-term forecasting supports efficient energy trading and scheduling, and long-term forecasting provides a foundation for strategic development and investment planning. By applying advanced AI techniques tailored to each forecasting horizon, the renewable energy sector can enhance reliability, reduce uncertainty and better integrate wind energy into the power grid.

10.2.2. *Wind turbine performance optimization*

Wind turbine performance optimization involves several advanced strategies such as the dynamic adjustment of turbine blade pitch, yaw angle and generator torque in response to changing wind conditions, in order to maximize energy generation. This not only maximizes energy capture but also minimizes mechanical stress, contributing to the longevity and efficiency of the turbines. Genetic Algorithms (GA) play a crucial role in improving the performance of wind farms and wind turbines.

Figure 10.2(a) and (b) shows a pictorial view of the dynamic control of yaw, blade pitch angle and wake control in wind turbines.

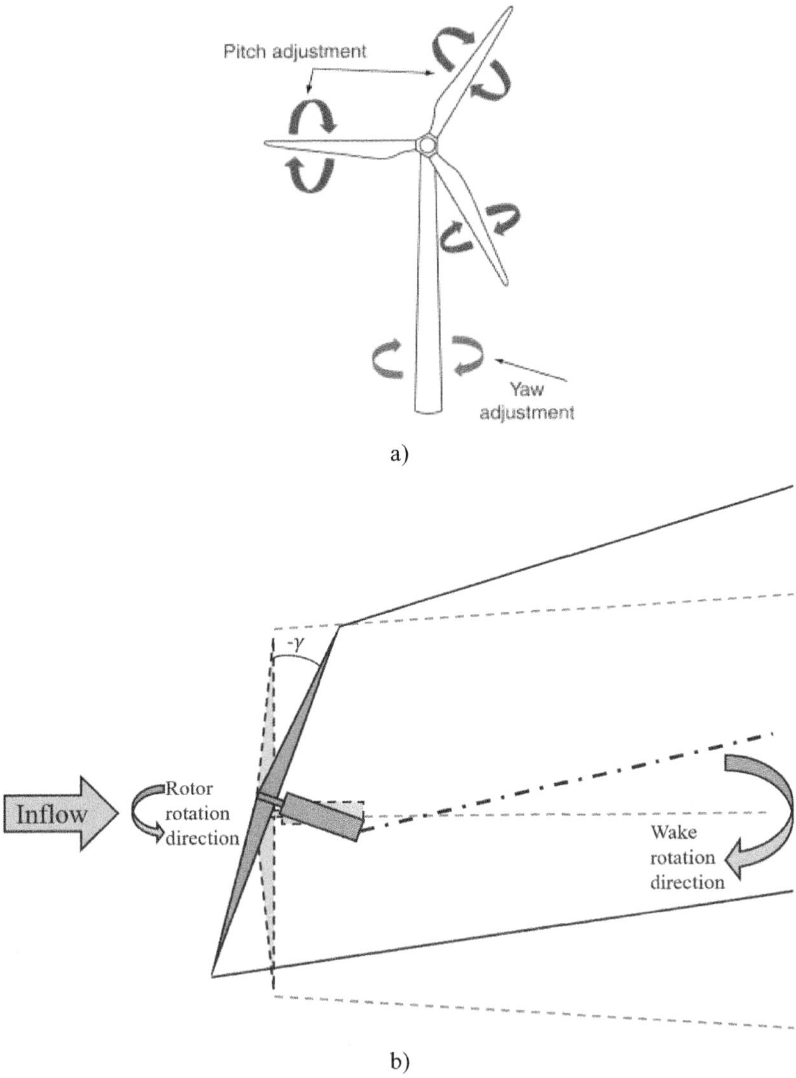

Figure 10.2. *(a) Blade pitch and yaw control. (b) Wake control. For a color version of this figure, see www.iste.co.uk/kumar/windpower2.zip*

10.2.2.1. *Blade pitch and yaw control*

The yaw and blade pitch angles are basics for the performance of a wind turbine. The yaw angle indicates the inclination of the entire turbine related to the direction of wind. The pitch angle of the blade refers to the angle at which the blades of the turbine are rotated with respect to the wind. At low wind speeds, the blades are adjusted to boost energy capture and in high winds, the blades are inclined to decrease the load on the turbine and avoid damage. This dynamic control ensures that the turbine operates within its optimal performance range under various wind conditions. The yaw control system adjusts the orientation of the entire turbine, ensuring it is always aligned with the wind direction. AI-based systems dynamically control both pitch and yaw to optimize performance and increase energy generation in real time, responding to the constantly changing wind conditions (Joshi et al. 2022).

10.2.2.2. *Wake-aware control*

In a wind farm, turbines are frequently affected by the wake of upstream turbines. This effect occurs when the wind slows down after passing through the blades of a turbine, creating turbulence that affects downstream turbines. By using advanced sensing and control algorithms using AI, the system can predict and minimize the impact of wake turbulence on downstream turbines. The goal is to minimize energy losses caused by wake interactions and increase the overall energy production in the wind farm. Wake-aware control strategies may also involve cooperative control among turbines, where each turbine adjusts its operation not only based on local wind conditions but also in consideration of the wake effects it is causing or experiencing from neighboring turbines. This optimization ensures that the turbines operate in a way that minimizes wake effects and improves the overall performance of the wind farm.

10.2.3. *Wind farm optimization*

Accurate wind resource assessments are essential for determining the viability and optimal placement of wind farms. Traditionally, the layout of wind turbines on a farm was based on simplified models and manual calculations that often failed to account for complex environmental and operational variables. With the integration of AI, particularly ML, evolutionary algorithms and deep reinforcement learning, engineers can now simulate and analyze millions of layout configurations quickly and with greater accuracy to provide a more precise assessment of wind energy potential at a given site (Yang et al. 2024). AI considers multiple dynamic factors including wind speed and direction variability, wake effects, terrain topography, turbine height, noise constraints, environmental impact and land-use restrictions to get the optimized design and layout of wind farms, ensuring that turbines are positioned where they will generate the most power. This enhances site selection and reduces the risk of poor performance due to unfavorable wind conditions. By processing historical weather data and predictive climate models, AI algorithms can determine the optimal positioning of each turbine to minimize energy losses caused by turbine interactions and maximize power generation. These systems can adapt layouts based on long-term operational goals, such as reducing maintenance costs or ensuring grid stability. For offshore wind farms, AI also factors in seabed conditions, water depth and cable routing to ensure structural feasibility and cost-efficiency. The result is a highly optimized wind farm design that improves the return on investment (ROI), extends the operational life of the turbines and supports more sustainable and reliable wind energy production.

GAs work by evaluating a range of potential turbine configurations and selecting the best ones through iterative processes. A population of possible wind farm layouts is created, and each layout is evaluated based on specific performance metrics such as energy production, wake interactions and land-use constraints. The most successful layouts are then selected for "reproduction" by combining features from different configurations. Over several iterations, the GA evolves increasingly optimized turbine arrangements. The objectives of the GA in wind farm optimization include maximizing energy production, minimizing the impact of wake effects and adhering to practical constraints such as land availability, maintenance access and environmental considerations. By considering all these factors, GAs can generate turbine layouts that outperform traditional methods, which often place turbines in a more arbitrary or grid-like pattern.

The combination of dynamic control of blade pitch and yaw angles, wake-aware control strategies and GAs leads to a highly optimized wind farm where turbines operate at their peak performance, energy generation is maximized and wake effects are minimized. AI-driven approaches allow for real-time adjustments and continuous optimization, while GAs offer a powerful method for designing optimal wind farm layouts. Together, these technologies are transforming the way wind farms are designed and operated, improving efficiency, reducing costs and contributing to the sustainability of renewable energy.

10.2.4. *Predictive maintenance and fault detection*

The most relevant use of AI in wind power systems is predictive maintenance, which involves the inspection of wind turbine components, especially critical parts such as the gearbox, rotor blades and bearings. Wind turbines are complex mechanical systems often installed in remote and harsh environments, where conventional maintenance strategies are expensive, time-consuming and inefficient. AI-driven predictive maintenance uses ML models and advanced analytics to anticipate whether the equipment will fail or require maintenance (Santiago et al. 2024). ML algorithms analyze historical and current data from the sensors on wind turbines such as vibration, temperature, pressure, speed, humidity and acoustic sensors to detect anomalies or early signs of wear and tear. AI software identifies trends and abnormalities from the sensor data in real time to anticipate the failure of the parts of the wind plant. Predictive models predict failures before they occur, enabling scheduled maintenance, thus reducing unplanned downtimes, extending equipment life and lowering operational costs. This technique extends the lifespan and reliability of a turbine by reducing unexpected failures and maintaining a steady power generation. AI-powered drone inspections can also be used to scan turbine blades and other hard-to-reach components, providing detailed images and identifying potential issues that require attention. The prediction of faults in wind machines was carried out using ML algorithms with past data collected from the wind farm in Prince Edward Island, Canada. Results show that the proposed algorithm predicted the failures of wind generators with an accuracy of 83% (Mammadov et al. 2021). According to the study made by the University of California, AI increases the lifespan of wind turbines by 10%. According to the study made by the Technical University of Munich, AI reduces the cost of wind energy by 15% (Datategy 2024). Figure 10.3 shows the fault detection model proposed by the authors.

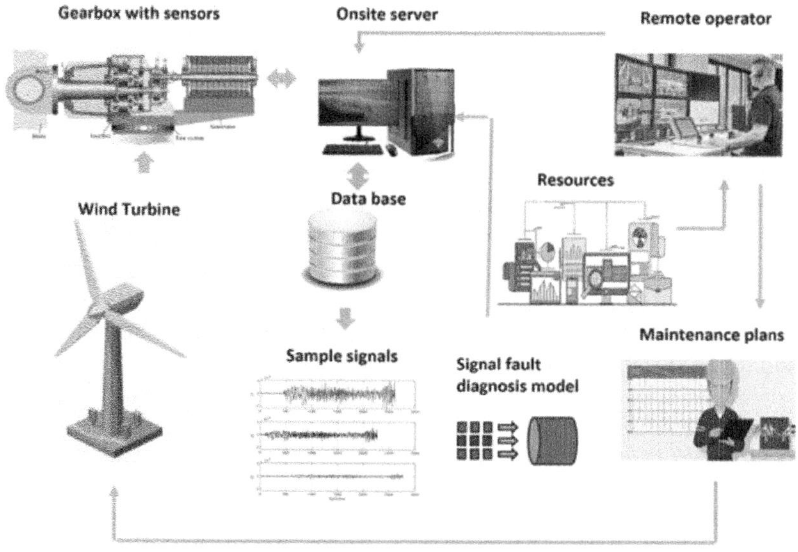

Figure 10.3. *Fault detection model used in wind turbines (Rubabaah 2023). For a color version of this figure, see www.iste.co.uk/kumar/windpower2.zip*

10.2.4.1. *Data collection*

The foundation of predictive maintenance lies in the continuous collection of operational data through embedded sensors installed on various parts of the wind turbine. These sensors sense key parameters such as gearbox vibration, generator temperature, rotor speed, nacelle orientation, oil pressure and environmental conditions such as humidity and wind speed. The collected data reflects the real-time health and performance of components and is transmitted to a central processing system, often cloud-based, using high-speed communication networks. This real-time data acquisition ensures that any deviations from normal operating behavior are promptly captured for further analysis.

10.2.4.2. *Data preprocessing*

Raw sensor data is often unstructured, noisy and contains missing or inconsistent values that can negatively impact model performance. Hence, preprocessing is an important step in making the data ready for analysis. During this stage, data is cleaned to remove outliers, noise or irrelevant records. Missing data points are imputed using interpolation or statistical techniques, and sensor readings may be normalized or standardized to maintain uniformity. Time-series data is often segmented into windows to capture short- and long-term trends. This ensures that the processed dataset is accurate, reliable and suitable for training ML algorithms.

10.2.4.3. *Feature engineering*

To improve the interpretability of ML models, engineers extract meaningful features from the raw sensor data. This is known as feature engineering, which converts complex data into a more compact and informative form.

For instance, vibration signals can be analyzed to derive statistical features such as mean, kurtosis, standard deviation or spectral features in the frequency domain using Fast Fourier Transform (FFT). Similarly, temperature trends over time, pressure deviations and rotor torque fluctuations can be summarized into numerical indicators that highlight early signs of wear or misalignment. Effective feature engineering helps models better understand the subtle differences between normal and failing conditions.

10.2.4.4. *Model training*

Once relevant features are extracted, the next step is training predictive models using historical labeled data, i.e. data with known outcomes such as failure events or normal operation. Supervised learning algorithms such as SVM, LSTM networks and gradient boosting are commonly used. These models learn to recognize patterns and correlations in the data that precede equipment failure. During training, the model is validated and fine-tuned to minimize false alarms while maximizing early detection accuracy. In some cases, unsupervised learning techniques are used when labeled failure data is limited, enabling anomaly detection based on deviations from learned "normal" behavior.

10.2.4.5. *Prediction and early detection*

After training, the model is deployed to continuously analyze the incoming sensor data and assess the health status of wind turbine components. If the system detects unusual patterns – such as a rise in gearbox temperature, an increase in generator vibration or declining oil pressure – it flags the event as a potential failure. The predictive model can provide a time window for when the issue might escalate, allowing operators to intervene proactively. These early warnings help in planning maintenance during non-peak wind periods, reducing emergency breakdowns and improving turbine availability. Ultimately, AI-driven predictive maintenance ensures more reliable, efficient and cost-effective operation of offshore and onshore wind farms.

Enabling real-time monitoring, early failure detection and proactive maintenance scheduling helps minimize downtime, reduce costs and improve operational efficiency. While challenges remain, the benefits of predictive maintenance, particularly in asset-intensive industries, are substantial, contributing to enhanced equipment reliability, extended asset life and safer working environments.

10.2.5. *Automation and remote monitoring*

Automation in wind farms enables intelligent, self-regulating control of turbines and their subsystems, ensuring operational efficiency, safety and minimal human intervention. Wind turbines are equipped with intelligent control systems that use embedded electronics and advanced microcontrollers to manage turbine operations autonomously. The system initiates turbine operation when wind speeds enter the range between 3 and 25 m/s and automatically shuts down in the event of extreme conditions such as severe vibrations, high wind speeds or critical mechanical faults, which ensures turbine safety and longevity. Another key component is yaw control, which keeps the rotor facing the wind direction by using wind vane sensors and yaw motors, thereby maximizing energy capture. Additionally, blade pitch control mechanisms continuously adjust the angle of the blades using electric or hydraulic actuators to maintain the optimal rotational speed and protect the turbine during high winds. Power regulation systems work in tandem to balance the power output according to fluctuating wind conditions and grid demands, ensuring grid stability. Emergency handling protocols are also embedded in the control logic, which automatically detect anomalies such as oil pressure drops or overheating and initiate protective actions such as braking or disconnecting the wind turbine from the grid.

SCADA systems serve as the nerve center of automation in wind farms, enabling centralized monitoring and control of all turbines from remote locations (Masoumi 2023). These systems collect and log real-time data on various operational parameters including rotor speed, generator temperature, voltage levels and wind speed, which is vital for performance optimization and predictive maintenance. SCADA systems also support alarm management by issuing alerts for abnormal conditions such as blade imbalance, overheating or electrical faults, allowing for quick diagnosis and response.

Automated fault handling is a critical feature in wind farm automation, leveraging predefined logic to respond to faults in real time without human intervention. These systems continuously analyze data from sensors embedded in turbine components such as gearboxes, generators and brake systems. When a fault is detected, such as overheating of the gearbox or excessive vibrations, the system automatically initiates corrective measures such as load reduction, controlled shutdown or engaging the braking system. This rapid response minimizes damage and prevents fault propagation. Over time, AI and ML algorithms enhance fault prediction accuracy by identifying subtle patterns or trends that precede failures, enabling proactive maintenance scheduling.

Remote monitoring supports continuous, real-time observation of turbine performance and health, significantly reducing the need for on-site personnel. Modern wind turbines are embedded with a wide range of advanced sensors that monitor critical parameters continuously. All this data is gathered in real time and transmitted to control centers or cloud-based platforms using robust communication technologies such as fiber-optic cables, satellite links or wireless systems such as LoRa and 5G. These communication channels are designed for high bandwidth and low latency to ensure that time-sensitive alerts and performance metrics are delivered without delay. This seamless integration enables fast decision-making and allows for continuous optimization of turbine operation from a remote location. Condition Monitoring Systems (CMS) are specialized platforms that analyze sensor data to evaluate the mechanical and electrical health of wind turbine components.

Digital twins represent a cutting-edge advancement in remote monitoring, offering a dynamic, virtual duplicate of a physical wind turbine. These digital models are continuously updated with instantaneous operational data from the site and integrated with simulation tools to predict how the turbine will respond under various conditions.

Operators can use digital twins to test different wind loads, extreme weather scenarios and control strategies to evaluate their impact before making changes to the actual turbine. Additionally, they facilitate advanced diagnostics and forecasting by comparing current behavior against expected performance, helping to identify inefficiencies, wear or anomalies. Over time, they evolve through ML and data analytics, making them increasingly accurate tools for strategic decision-making, maintenance planning and operational excellence in offshore wind farms.

10.2.6. *Autonomous drones for inspection and monitoring*

Wind turbines are more prone to various anomalies and defects that can significantly affect their operation and hence shoot up the maintenance costs. The most common issue is blade erosion caused by rain and airborne particles, which decreases aerodynamic efficiency leading to structural damage. Sometimes, defects in turbine blades are neglected during production, creating long-term risks to safety and the operation of the turbine. Common flaws may be micro-cracks and voids inside composite materials that can weaken the blade over time and cause premature failure and costly repairs or reduced energy output. According to a report by the American Bureau of Safety and Environmental Enforcement, 85% of blade failures are due to poor maintenance, which highlights the significance of regular inspections and proper maintenance in order to prevent costly downtime and ensure the efficient operation of turbines over their lifespan.

Besides these anomalies in the structures, the location and height of wind turbines make inspections inherently dangerous. Inspectors perform most interventions manually, assessing vibrating structures and high-speed spinning blades using rope access techniques. A report by the American Bureau of Safety and Environmental Enforcement states that nearly 249 accidents resulted in 161 fatalities related to manual inspections of turbines in the past few decades. AI-powered drones with advanced ML and image technologies are modifying the game in the inspection of wind farm. Drones take thermal imaging and high-resolution photographs using advanced sensors and cameras. ML identifies defective components, corrosion and cracks through images which are then assessed by AI. Thus, these drones help to ensure optimal turbine performance by scheduling and carrying out timely inspections, minimizing human exposure and lowering maintenance costs.

10.2.7. *Bird avoidance systems in wind farms*

AI-driven bird avoidance systems are designed to reduce the risk of bird collisions with wind turbines, a significant concern for wildlife conservation, particularly in regions with high bird activity (Ojija et al. 2025). These systems use real-time data collected from sensors such as radar, infrared cameras and optical systems to track bird movements. ML algorithms analyze the data, identifying flight patterns and predicting potential collisions. In order to avoid the possibility of bird strikes, the system can then automatically modify the turbine operations, such as stopping or slowing the blades. This AI-enabled approach allows wind farms to operate with minimal human intervention while ensuring wildlife protection and compliance with environmental regulations.

10.2.8. *Design of turbine blades*

AI-designed turbine blades represent a cutting-edge advancement in wind energy technology, where AI is used to optimize the shape, size and material of wind turbine blades based on real-time data. Traditionally, blade design relied on aerodynamic simulations and iterative testing, which were time-consuming and limited in their ability to explore complex design possibilities. A study by the National Renewable Energy Laboratory (NREL) indicated that AI can enhance wind turbine efficiency by 20%. Using advanced ML algorithms, AI models analyze large datasets such as wind flow patterns, material properties, structural stress behavior and historical performance data to create blade designs that increase energy capture while reducing noise, weight and material fatigue (Jayabalan et al. 2017). With ML and generative design algorithms, engineers can now simulate various conditions to test thousands of blade configurations and predict their performance under specific circumstances. Techniques such as reinforcement learning and neural networks are used to learn the relationship between blade geometry and its aerodynamic efficiency under varying wind conditions. As a result, AI-designed blades are often more curved, twisted or even asymmetric compared to conventional blades, enabling better adaptability to turbulent offshore environments or low-wind regions. This approach not only boosts the efficiency and lifespan of turbines but also reduces manufacturing costs by identifying material-saving opportunities without compromising strength or performance.

10.2.9. *Grid integration and energy storage optimization*

One of the main challenges of wind power is its integration with existing grids since the variability of winds causes instability. AI-powered algorithms help in making smooth grid integration a reality by optimal charging and discharging of energy storage and wind power distribution, ensuring minimal power loss and grid congestion. By forecasting both wind production and energy demand, AI can help optimize when to store excess energy or release it to the grid, thus ensuring better alignment between generation and consumption. This is especially important as wind energy is becoming a larger share of the overall energy mix.

10.3. Case studies

10.3.1. *AI-driven predictive maintenance*

Siemens Gamesa, a global leader in wind turbine manufacturing, has integrated AI into its operations to implement predictive maintenance solutions. Using advanced analytics embedded within their SCADA and CMS, the company collects and processes real-time data from numerous sensors installed on wind turbines (Siemens Energy 2025). These sensors monitor critical variables such as gearbox vibration, blade movement, oil pressure, rotor temperature and structural stress. AI models, particularly those trained with historical data, are used to detect abnormal trends and predict potential faults long before they occur. For instance, subtle changes in vibration patterns may indicate bearing wear, while fluctuations in oil temperature could signal lubrication issues. This proactive approach allows Siemens Gamesa to schedule maintenance during optimal windows, avoid catastrophic failures and reduce operational costs. The AI-driven system improves turbine availability, reduces downtime and extends the service life of mechanical components by ensuring maintenance is carried out based on the equipment's condition rather than on a fixed schedule.

AES produces gigawatts of renewable energy by running several wind turbine farms. As they are operated in harsh environments, many parts of the wind turbines experience significant vibration resulting in their wear and tear. Maintenance teams required expensive and specialized cranes for carrying out the inspection work. They wasted money and time in sending vehicles for repair work only to discover that they had forgotten to bring the right parts. An initiative was taken by the AES data science team to forecast failures on dozens of wind turbine components (AES Energy 2025). They developed a dozen AI models with the available data from the wind turbine manufacturers and integrated with H_2O AI Cloud. Scenario analysis is also incorporated in the final outputs to give key intelligence to the model. This initiative immediately earned a saving of millions of dollars and steady power delivery. At present, the AES team has 35 AI models to help in predicting failures across their fleet of wind turbines. They are now planning to include images from drones to recognize cracks on the towers and blades.

10.3.2. *AI-based wind farm layout optimization by DeepMind*

DeepMind, Alphabet's AI research arm, partnered with Google to solve one of the major challenges in wind energy: unpredictability (DeepMind 2025). By using deep learning models, DeepMind created a system to forecast wind power for 36 hours in advance. The AI system analyzes a combination of real-time weather data, historical turbine performance and market conditions to predict how much electricity a wind farm will generate. This prediction capability allows operators to plan energy delivery to the grid more accurately, improving both the reliability and economic value of wind power. Additionally, DeepMind used reinforcement learning algorithms to explore optimal turbine layouts that minimize wake losses due to the aerodynamic turbulence caused by upstream turbines affecting downstream ones. By simulating thousands of layout scenarios, the AI identifies the best turbine placements to maximize the overall energy capture. These innovations proved that Google's wind energy output increased by 20–30%, justifying the use of AI in bridging the gap between the variability of renewable energy and grid demand.

10.3.3. *Vestas AI-powered smart turbine fleet management*

Vestas has embraced AI to revolutionize its wind turbine fleet management, covering thousands of onshore and offshore installations globally. Using an AI-based platform, Vestas collects massive volumes of real-time data from turbine sensors, capturing metrics such as wind speed, blade angle, rotor torque, temperature and vibration (Vestas 2025). AI algorithms analyze this data continuously to optimize turbine performance. For example, if one turbine in a region is consistently outperforming others under similar wind conditions, the AI system can detect the control strategy used and replicate it across the fleet. The platform also provides early warnings about potential component failures, enabling predictive maintenance and minimizing unexpected downtimes. Moreover, by integrating environmental data, Vestas adjusts turbine configurations dynamically to suit prevailing wind patterns and grid requirements. This intelligent, AI-driven management strategy enhances power output, reduces wear on equipment, improves operational efficiency and supports more informed, strategic decision-making across the company's global wind assets.

10.3.4. *GE Renewable Energy and AI for optimizing turbine performance*

GE Renewable Energy has created Predix, an IoT platform that uses AI to enhance the reliability and performance of wind turbines. Figure 10.4 shows the Predix Platform based on the IoT.

Predix collects and analyzes sensor data from every turbine component, including rotor blades, gearboxes, generators and inverters. Using ML algorithms, the platform identifies real-time opportunities to improve energy capture, such as adjusting the blade pitch or yaw angle in response to fluctuating wind conditions. These continuous adjustments allow each turbine to operate at peak efficiency under all conditions. Additionally, Predix's predictive maintenance capabilities monitor for anomalies that could indicate impending failures such as overheating components or misaligned shafts, allowing service teams to intervene before breakdowns occur. The platform also facilitates grid integration by forecasting output and demand, thus helping maintain grid stability. With AI, GE enables turbines to be more self-aware and adaptive, reducing unplanned maintenance, improving energy efficiency and lowering lifecycle costs across wind farms (GE Vernova 2025a).

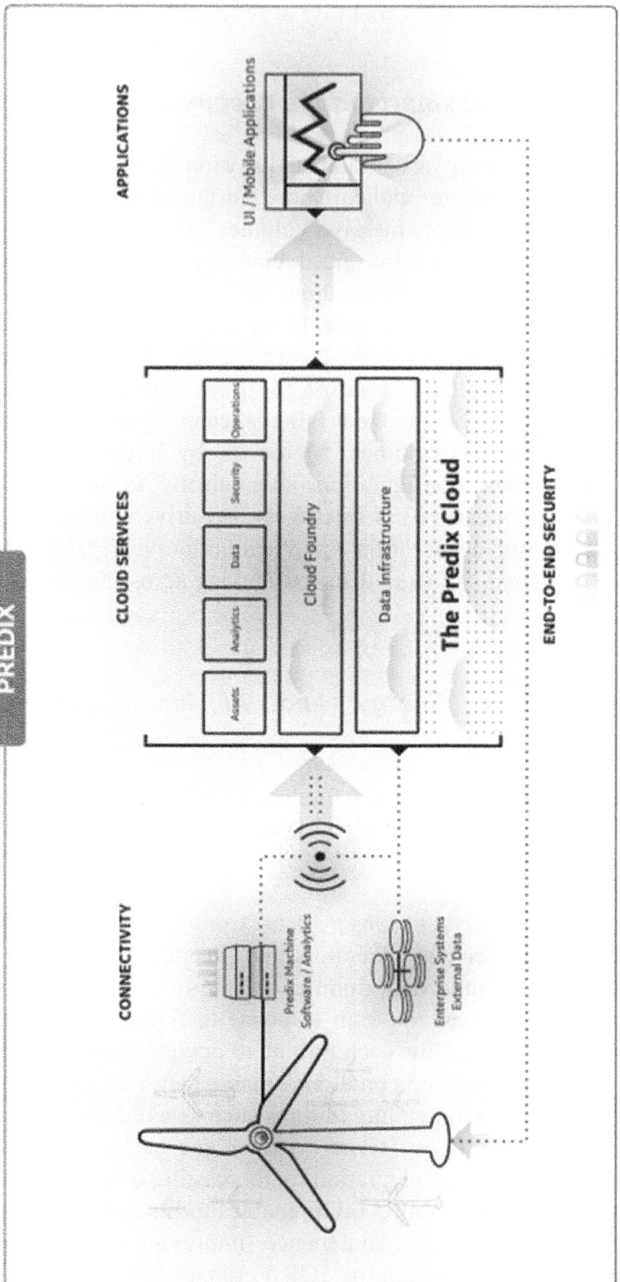

Figure 10.4. *Predix Cloud (GE Vernova 2025b). For a color version of this figure, see www.iste.co.uk/kumar/windpower2.zip*

10.3.5. *AI-driven bird avoidance systems – the IdentiFlight system*

IdentiFlight is an AI-powered visual recognition system that combines high-precision optics and ML to detect, classify and track flying birds in real time (Duke Energy 2025). The system uses high-resolution stereoscopic cameras mounted on towers that scan the surrounding airspace for avian activity. Once a bird is detected, the AI software instantly identifies the species, particularly focusing on high-risk species such as eagles or vultures using a trained image recognition model (IdentiFlight 2025). This classification happens within seconds, allowing the system to assess whether the bird's flight path poses a collision risk with any turbine in the vicinity. If there is any threat, the system sends signals to the control system of turbine to shut down or slow down the particular turbine until the bird passes safely. This targeted and temporary action ensures minimal disruption to energy production while providing effective protection to wildlife. Field tests and long-term observations have demonstrated a significant reduction in eagle fatalities by as much as 82% in some wind farms when the IdentiFlight system is in use compared to control sites without such technology. Data collected by IdentiFlight over time helps in habitat mapping and migration analysis, providing valuable insights for environmental assessments and future wind farm planning. This proactive approach not only ensures regulatory compliance but also strengthens the industry's commitment to sustainability and biodiversity. AI-driven bird avoidance systems such as IdentiFlight represent a significant step towards harmonizing renewable energy development with wildlife conservation. By leveraging AI's speed, accuracy and learning capabilities, wind farms can mitigate ecological risks without compromising operational efficiency.

10.4. Challenges of AI in the wind industry

Several challenges hinder the effective implementation of AI in wind farms, including data quality and availability, integrating AI with existing systems, and regulatory and ethical considerations (Clir 2025). Other challenges include the complexity of the system, cyberattacks and the question of liability in the case of accidents. Additionally, understanding how to leverage new technologies effectively, collaborating across the industry and ensuring well-labeled data are key to maximizing AI's potential in wind energy.

10.4.1. *Data quality and availability*

Each turbine generates more than 500 data points every 10 minutes, meaning that wind farm operators have large volumes of data at their disposal. AI systems require vast amounts of high-quality data for training and decision-making. Collecting and cleaning data from various sources within a wind farm, such as turbine sensors, weather stations and operational logs, can be complex. Poor-quality or insufficient data may lead to imperfect AI models and reduced efficiency. Wrong conclusions may be drawn from the tests that are conducted in the laboratory which differ from the conditions of the field. For example, an LSTM model for identifying failure of a bearing attained 99.5% accuracy in lab tests, but in the field the model misidentified normal vibrations as failures due to noise in the environment.

10.4.2. *Integration with existing systems*

Integrating AI solutions with the existing wind farm infrastructure can be a complex and costly process. Connecting AI algorithms to existing SCADA systems and other operational tools can require significant engineering and programming efforts. A lack of interoperability or incompatibility can hamper the implementation and integration of AI.

10.4.3. *Regulatory and ethical considerations*

There are issues such as data privacy, security and the transparency of decision-making processes due to the use of AI. Ensuring data security and privacy when collecting and analyzing sensitive operational data from wind turbines is crucial. Failure to address these issues can lead to ethical and legal challenges, hindering the adoption of AI. When AI is used in decision-making, it can lead to ethical dilemmas, particularly related to the choices which have an impact on the environment or human lives. In the event of accidents caused by AI decisions, it is difficult to find out whether the developer, manufacturer or user is responsible. This ambiguity may result in a lack of accountability and lower inspiration to develop efficient and safe systems.

10.4.4. *Understanding and leveraging AI*

It is difficult to manage or understand ML, AI and big data as they are very complex and sophisticated, which may cause issues in optimizing and maintaining the system. The best way to use all these techniques in wind power is still not well understood. Unfortunately, wind power developers need to waste a significant amount of time and money in carrying out poor solutions. To leverage AI technologies efficiently, they need to acknowledge that AI is just a tool and not a solution to the problems.

10.4.5. *Collaboration and data sharing*

The wind energy industry often involves multiple stakeholders (operators, owners, OEMs), which can make data sharing and collaboration challenging. Sharing anonymized data from different wind farms or turbine manufacturers can enable the creation of more accurate and robust AI models. OEMs consider turbine data as proprietary information and are hesitant to share data. Meanwhile, owners cannot access the data from their own portfolio to supervise asset performance. This hinders their ability to get benchmarking data and gain understanding in energy production, chances for optimization, reporting capabilities, etc.

10.4.6. *Well-labeled data*

Data from wind farms is often not well-labeled or standardized, making it difficult to train AI models effectively. Standardizing data formats, ensuring consistent labeling of data streams and creating comprehensive metadata can enhance the data quality used to form AI models. With a vast amount of contextual data and operations available, it is essential that data must be standardized, clean and well-labeled. Unfortunately, this is not the case when coming across wind portfolio data. Poorly labeled data can lead to inaccurate predictions and a reduced performance of AI-driven systems. Without standardized and clean data, it is impossible to derive insights or information for informed decisions.

10.4.7. *Algorithmic bias and hallucinations*

AI systems do not have the capability to understand the context or circumstances and hence produce "hallucinations", which may result in fabricated or inaccurate results. For example, in wind turbines, fault detection models can concentrate on minor problems such as blade cracks, but miss critical but rare failures such as wear of the gearbox or bearing.

10.4.8. *Cyberattacks*

AI systems are more prone to cyberattacks, which may result in an outage or physical damage to wind turbines. Attackers can make use of vulnerabilities in AI systems to take over the control of wind plants. For example, a catastrophic failure may occur if manipulated sensor data could trick AI into ignoring bearing failures or blade cracks. This is why cyber-protection systems are needed. An example of a cybersecure control system of wind turbine makes use of an $H\infty$ controller with an SVM algorithm to identify data injection attacks or communication attacks.

10.5. Conclusion

As the use of wind energy increases to meet the global demand, the ability of industry to traverse safety, environmental challenges and complex maintenance is crucial. By leveraging digital transformation, led by data-driven and AI tools, the wind industry is facing a paradigm shift to make wise decisions, enhance maintenance strategies and protect both the environment and human beings. From reducing bird collisions to optimizing turbine blades, inspecting critical components and improving wind resource assessments, AI is enabling wind farms to operate more efficiently and safely. With greater forecasting accuracy, AI allows grid managers to balance demand and supply more efficiently, with less need for conventional fuels as backup. With the world focusing on sustainable development goals, AI is the game-changer that can guarantee increased reliability and efficiency of wind power. Better balance among power consumption and generation can be achieved by integrating AI into smart grids and energy storage systems, thereby improving the power quality and minimizing the carbon traces of electricity. Dangers related to manual inspections have been reduced by drone-based inspections in conjunction with computer vision techniques, particularly in high-altitude scenarios or offshore wind farms. Upcoming research could incorporate more trusted AI models that foster confidence with the public, regulators and engineers investigating the AI-driven solutions in emerging power system architectures such as microgrids, deregulated energy markets, strong cybersecurity measures and clear regulatory guidelines. If these issues are addressed carefully, AI-driven solutions can assist the world energy sector in meeting climate targets, improving grid resilience and promoting a sustainable future.

10.6. References

AES Energy (2025). AES transforms its energy business with AI and H_2O.ai [Online]. Available at: https://h2o.ai/case-studies/aes-transforms-energy-business-with-ai-and-h2o.

Borunda, M., Ramírez, A., Garduno, R., García-Beltrán, C., Mijarez, R. (2023). Enhancing long-term wind power forecasting by using an intelligent statistical treatment for wind resource data. *Energies*, 16(23), 7915.

Chen, Q. and Folly, K.A. (2018). Comparison of three methods for short-term wind power forecasting. In *2018 IEEE International Joint Conference on Neural Networks (IJCNN)*, 1–8.

Chen, X., Eder, M.A., Shihavuddin, A., Zheng, D. (2021). A human-cyber-physical system towards intelligent wind turbine operation and maintenance. *Sustainability*, 13, 561.

Clir (2025). AI for wind energy [Online]. Available at: https://www.clir.eco/blog/ai-for-wind-energy [Accessed 13 June 2025].

Datategy (2024). Wind turbine optimization with AI [Online]. Available at: https://www.datategy.net/2024/01/17/wind-turbine-optimization-with-ai/.

DeepMind (2025). Machine learning can boost the value of wind energy [Online]. Available at: https://deepmind.google/discover/blog/machine-learning-can-boost-the-value-of-wind-energy/ [Accessed 13 June 2025].

Duke Energy (2025). How Duke Energy is using technology to save eagles at a Wyoming wind site [Online]. Available at: https://illumination.duke-energy.com/articles/how-duke-energy-is-using-technology-to-save-eagles-at-a-wyoming-wind-site [Accessed 13 June 2025].

Gasch, R. and Twele, J. (eds) (2011). Introduction to wind energy. In *Wind Power Plants: Fundamentals, Design, Construction and Operation*. Springer, Berlin, Heidelberg.

GE Vernova (2025a). GE using AI/ML to reduce wind turbine logistics and installation costs [Online]. Available at: https://www.gevernova.com/news/press-releases/ge-using-aiml-to-reduce-wind-turbine-logistics-and-installation-costs [Accessed 13 June 2025].

GE Vernova (2025b). Predix platforms [Online]. Available at: https://www.gevernova.com/software/documentation/predix%20platforms/ [Accessed 13 June 2025].

Ghosh, M. and Arunachalam, T. (2021). Introduction to artificial intelligence. In *Artificial Intelligence for Information Management: A Healthcare Perspective*, Srinivasa, K.G., Siddesh, G.M., Mani Sekhar, S.R. (eds). Springer, New York.

IdentiFlight (2025). How it works [Online]. Available at: https://www.identiflight.com/howitworks?utm_source=chatgpt.com#gsc.tab=0 [Accessed 13 June 2025].

Jamii, J., Trabelsi, M., Mansouri, M., Kouadri, A., Mimouni, M.F., Nounou, M. (2024). Medium-term wind power forecasting using reduced principal component analysis based random forest model. *Wind Engineering*, 48(4), 597–616.

Jayabalan, J., Yildirim, D., Kim, D., Samui, P., (2017). Design optimization of a wind turbine using artificial intelligence. In *Mathematical Concepts and Applications in Mechanical Engineering and Mechatronics*, Ram, M. and Davim, J.P. (eds). IGI Global, Hershey, PA.

Joshi, M.P., Mallikarjunaiah, S., Rao, S., Umarani, B., KarthiKumar, K., Subramaniam, S., (2022). Advanced ai based controller for wind turbine pitch control. In *2022 IEEE Sixth International Conference on I-SMAC (IoT in Social, Mobile, Analytics and Cloud) (I-SMAC 2022)*, Dharan, 951–955.

Mammadov, E., Farrokhabadi, M., Cañizares, C.A., (2021). AI-enabled predictive maintenance of wind generators. In *2021 IEEE PES Innovative Smart Grid Technologies Europe (ISGT Europe)*, 1–5.

Masoumi, M. (2023). Machine learning solutions for offshore wind farms: A review of applications and impacts. *Journal of Marine Science and Engineering*, 11(10), 1855.

Moshtaghi, P., Hajialigol, N., Rafiei, B. (2025). A comprehensive review of artificial intelligence applications in wind energy power generation. *Sustainable Futures*, 9, 1–17.

Ng, C. and Ran, L. (eds) (2016). Introduction to offshore wind energy. In *Offshore Wind Farms: Technologies, Design and Operation*. Woodhead Publishing, Duxford, 3–8.

Ojija, F., Ogwu, M.C., Ally, J., John, J.P., Stephano, A., Felix, N., Tekka, R. (2025). Artificial intelligence-driven solutions for mitigating human–wildlife conflict in biodiversity hotspots. *Science Progress*, 108(4), 00368504251394584.

Rababaah, A.R. (2023). Comparative study of deep learning models versus machine learning models for wind turbine intelligent health diagnosis systems. *Arabian Journal for Science and Engineering*, 48(8), 10875–10899.

Santiago, R.A.D.F., Barbosa, N.B., Mergulhão, H.G., Carvalho, T.F.D., Santos, A.A.B., Medrado, R.C., Filho, J.B.D.M., Pinheiro, O.R., Nascimento, E.G.S. (2024). Data-driven models applied to predictive and prescriptive maintenance of wind turbine: A systematic review of approaches based on failure detection, diagnosis, and prognosis. *Energies*, 17(5), 1010.

Shabbir, N., Kütt, L., Jawad, M., Husev, O., Rehman, A.U., Gardezi, A.A., Shafiq, M., Choi, J.G. (2022). Short-term wind energy forecasting using deep learning-based predictive analytics. *Computers, Materials & Continua*, 72(1), 1017–1033.

Siemens Energy (2025). Diagnostic, monitoring & protection services [Online]. Available at: https://www.siemens-energy.com/global/en/home/products-services/service/diagnostic-monitoring-protection.html [Accessed 13 June 2025].

Stadtmann, F., Rasheed, A., Kvamsdal, T., Johannessen, K.A., San, O., Kölle, K., Tande, J.O., Barstad, I., Benhamou, A., Brathaug, T. et al. (2023). Digital twins in wind energy: Emerging technologies and industry-informed future directions. *IEEE Access*, 11, 110762–110795.

Sun, Y. and Han, W. (2025). A review of enhancing wind power with AI: Applications, economic implications, and green innovations. *Digital Economy and Sustainable Development*, 3(1), 11.

Vestas (2025). Fleet optimisation [Online]. Available at: https://www.vestas.com/en/energy-solutions/service/fleet-optimisation [Accessed 13 June 2025].

Yang, S., Deng, X., Yang, K. (2024). Machine-learning-based wind farm optimization through layout design and yaw control. *Renewable Energy*, 224, 120161.

11

Generative AI for Energy Consumption Behavior Analysis

This chapter explores the role of generative artificial intelligence (AI) in analyzing and modeling energy consumption behavior within modern smart grids. By integrating advanced machine learning and simulation methods, generative AI models offer novel approaches for enhancing demand forecasting, privacy preservation and energy system optimization. The discussion aims to bridge data science and sustainable energy management by showcasing the transformative potential of these emerging technologies.

The increasing complexity of energy consumption patterns in smart grids and distributed energy systems requires advanced analytical tools capable of handling stochasticity, uncertainty and nonlinear dependencies. generative artificial intelligence (AI) has emerged as a powerful paradigm for modeling consumer behavior by generating synthetic yet realistic energy usage profiles.

This chapter determines the methodological underpinnings of generative approaches, including variational autoencoders (VAEs), generative adversarial networks (GANs) and diffusion-based models, in the context of energy demand forecasting, anomaly detection and behavioral clustering. The main emphasis is placed on the integration of generative AI with traditional machine learning and optimization frameworks to improve load prediction accuracy, support demand-side management and evaluate policy interventions under diverse simulated scenarios. Empirical case studies highlight the robustness of generative models in overcoming data scarcity, reducing model bias and capturing fine-grained behavioral heterogeneity.

This chapter starts with a discussion on computational challenges, data fidelity of synthetic data and emerging research frontiers in generative modeling for energy informatics.

Chapter written by S. VANSHIKA and Neetu RANI.

11.1. Introduction

The global energy sector is currently undergoing a fundamental transformation. Driven by the critical needs of combating climate change and integrating volatile renewable energy sources, the traditional, centralized power grid is evolving into a more flexible, smarter and decentralized system. In this new paradigm, understanding and influencing how people use energy is no longer a secondary issue (Bu et al. 2021). It is a core requirement for grid stability, efficiency and sustainability. For many years, energy providers and researchers have depended on classic statistical and machine learning methods to analyze consumption trends. While these techniques have their uses, they often struggle to capture the full complexity, unpredictability and subtle behavioral drivers that influence human energy use. Traditional analytical methods in energy analysis, such as statistical models and conventional machine learning, have proven effective at predicting what is likely to happen. However, they are often limited in their ability to create novel and realistic scenarios or provide a comprehensive understanding of the complex factors that influence energy usage. These methods are designed to classify or predict based on existing data, but they lack the creative capacity to generate new data or explore "what if" situations.

This limitation is precisely where generative artificial intelligence (AI) offers a significant advancement. Unlike discriminative AI models – which are built to categorize data or forecast outcomes (e.g. identifying a customer as a "high user") – generative models are designed to produce new data that replicates the characteristics of their training data. In the context of energy analysis, this allows them to learn the fundamental patterns of consumption and then generate completely new, yet highly plausible, load profiles. This capability opens up a wide array of powerful applications, from creating synthetic datasets that protect consumer privacy to generating believable future scenarios for more resilient grid planning.

This chapter will explore the use of generative AI in analyzing energy consumption. We will first examine the fundamental concepts of key generative models, such as generative adversarial networks (GANs) and variational autoencoders (VAEs), and show how they can be specifically applied to the time-series data from smart meters (Ji et al. 2025). The discussion will then shift to practical applications, including the creation of synthetic data to protect consumer privacy, advanced forecasting, behavioral clustering and the development of personalized energy recommendations. Finally, we will address the implementation challenges, ethical considerations and the future outlook for this rapidly evolving field. By the end, you will have a clear grasp of both the theoretical underpinnings of generative

AI and its potential to revolutionize how we understand, manage and shape our energy future.

11.2. Fundamentals of generative AI for energy consumption behavior

Generative AI provides powerful tools for modeling and simulating energy consumption behavior. Unlike traditional predictive models that only forecast future values, generative models learn the underlying distribution of consumption data and can produce synthetic yet realistic usage patterns. This capability is critical in the energy sector, where data scarcity, privacy concerns and behavioral uncertainty often limit robust analysis.

According to the generative AI model taxonomy, the key families of generative models and their applications to energy consumption are as follows.

11.2.1. *Generative adversarial networks (GANs)*

GANs operate on a collaborative model where a generator creates artificial data and a discriminator assesses its realism. This dynamic makes GANs highly useful in energy consumption analysis, especially when historical data is scarce. They can produce lifelike household or industrial load curves, providing utilities and researchers with richer datasets for building more accurate forecasting models (Lin et al. 2024).

Advanced GAN architectures have proven particularly effective. Deep Convolutional GANs (DCGANs) and Gaussian Adaptive Adversarial Networks (GAANs), for example, are used to create detailed time-series profiles of daily electricity use, offering key insights into how consumers use power. The Cycle GAN architecture takes this a step further, enabling transfer learning to adapt data from one context to another, such as converting summer demand patterns to their winter equivalents (Liu et al. 2023). Additionally, Conditional GANs (cGANs) can generate load profiles influenced by specific variables such as weather, pricing or individual appliance use. This capability is valuable for scenario analysis and for testing new policies in modern energy grids.

Beyond generating data, GANs are also used for anomaly detection, helping identify unusual consumption patterns that could signal energy theft or faulty meters. Their capacity to simulate diverse demand scenarios can also support demand response (DR) strategies, allowing grid operators to predict how consumers might react to variable pricing or increased renewable energy sources. As energy systems become more reliant on data, the versatility of GANs makes them a fundamental technology for advancing energy analytics and developing future smart grid solutions.

11.2.2. *Variational autoencoders (VAEs)*

VAEs are a powerful tool for modeling the variability in energy consumption. They operate by compressing data into a latent representation, then using that compressed form to generate new, realistic instances. This makes them highly effective for simulating smooth, natural variations in household behavior, which in turn helps create personalized demand-side management strategies (Ma and Yeung 2024).

Different VAE architectures offer specialized capabilities:

– β-VAEs can help researchers isolate different behavioral drivers, such as separating the impact of a home's occupancy from the efficiency of its appliances. This provides a more detailed understanding of demand patterns.

– Conditional VAEs can generate specific consumption profiles based on particular conditions, such as weekday versus weekend usage or urban versus rural settings, enabling scenario-based modeling.

– Hierarchical VAEs can capture multi-level patterns, allowing for the aggregation of individual household demand into broader neighborhood or city-level trends.

Beyond generating data, VAEs are also useful for anomaly detection. A high reconstruction error can signal irregular consumption patterns, potentially indicating a faulty meter or unauthorized usage. They are also instrumental in probabilistic forecasting, where instead of a single prediction, they provide a distribution of potential demand. This is crucial for managing the uncertainty that comes with integrating renewable energy sources (Razghandi et al. 2023).

By modeling the subtle consumption variations among diverse consumer groups, VAEs can help shape new adaptive tariff policies, personalized energy recommendations and more resilient DR programs. Their blend of interpretability and generative power makes VAEs a versatile asset for developing smarter energy systems.

11.2.3. *Flow-based models*

Generative models such as real NVP and glow, known as flow-based approaches, function by learning invertible mappings. This means they can transform a simple probability distribution, such as a Gaussian, into a more complex distribution that mirrors real-world data, and vice versa. A key advantage of these models is their ability to perform exact probability estimation, avoiding the approximations used by other generative methods. This makes them especially valuable for tasks in the energy sector where statistical precision is essential.

In the energy domain, these models are excellent for grid monitoring and anomaly detection. They can precisely pinpoint unusual consumption spikes or deviations from expected patterns. For instance, flow-based models have been used to identify energy theft, irregular household usage and operational errors in industrial systems. Furthermore, their capacity to create subtle yet realistic variations in industrial energy demand allows for stress-testing of load-balancing algorithms under a wide range of conditions.

Beyond detecting anomalies, flow-based models also support probabilistic forecasting by providing not just a single prediction but a complete probability distribution of future consumption (Sánchez-Mompó et al. 2025). This is crucial for managing the uncertainty associated with grids that rely heavily on renewable energy. These models can also effectively capture the relationships between different energy variables, such as electricity, gas and heat demand, which supports integrated energy system planning. Their interpretability, a result of their invertible nature, makes them appealing for regulated industries where the transparency of AI-driven decisions is mandatory. As energy systems become more intricate, flow-based models serve as precise, explainable instruments for analyzing consumption behavior and ensuring grid reliability.

11.2.4. *Transformer-based models*

Originally developed for natural language processing, transformers are a powerful class of generative models that are being increasingly applied to energy systems. Their ability to handle sequential data and capture long-range dependencies makes them perfect for modeling temporal patterns in energy consumption. By incorporating external factors such as weather, dynamic pricing or new policies, transformers can provide highly adaptive and context-aware demand forecasts.

For instance, models similar to GPT-3 have been used to create narrative explanations of consumption trends, giving human-readable insights alongside numerical predictions. While models such as DALL·E 2 or stable diffusion were built for images, they can be adapted to visualize synthetic demand scenarios or consumption heatmaps. This helps energy planners explore potential future scenarios. Additionally, BERT-based encoders can be used to enrich forecasting models by embedding information about customer type, regulatory context or market shifts, thereby improving accuracy (Sun et al. 2023).

Beyond just forecasting, transformers enable multi-modal analysis, allowing for a richer understanding of consumption by combining text, numerical data and visuals. They are also being explored in reinforcement learning pipelines, where their generative predictions can guide adaptive demand–response strategies. Their scalability and pretraining capabilities make them ideal for transfer learning; models trained on large datasets from one region or sector can be fine-tuned for use in another. As energy systems become more complex, transformers provide a flexible, scalable and explainable foundation for advancing generative AI in energy consumption analysis. Taxonomy of generative AI is shown in Figure 11.1.

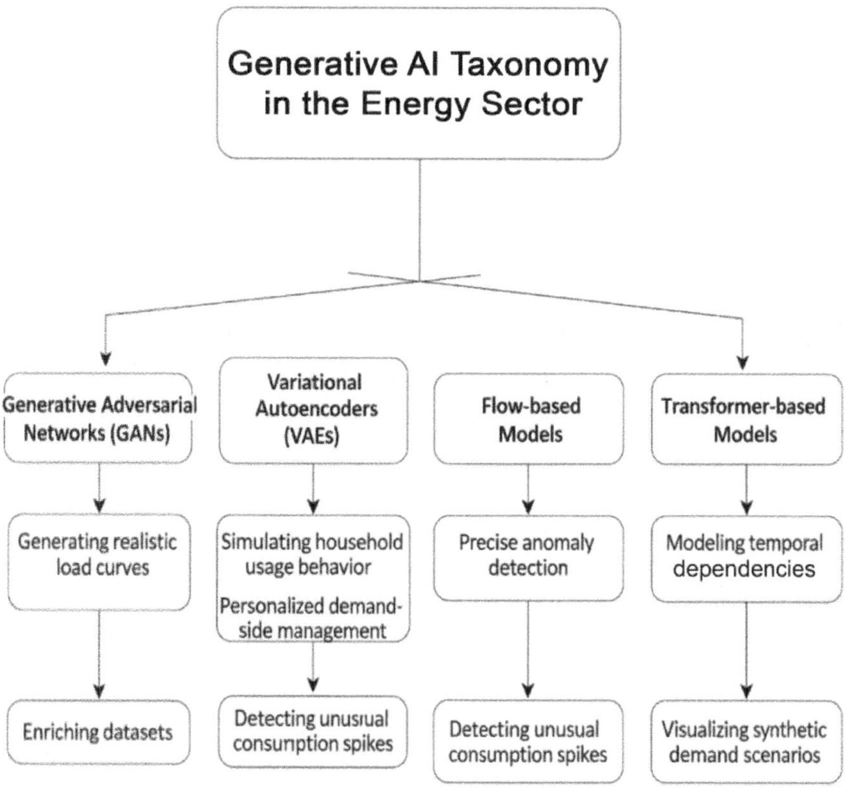

Figure 11.1. *Taxonomy of generative AI in the energy sector*

11.3. Data in energy behavior analysis

The fuel for any generative AI model is data. In the context of energy consumption, the primary data source is advanced metering infrastructure (AMI), commonly known as smart meter data (Su et al. 2023).

Some of the characteristics of smart meter data are shown in Figure 11.2.

11.3.1. *Characteristics of smart meter data*

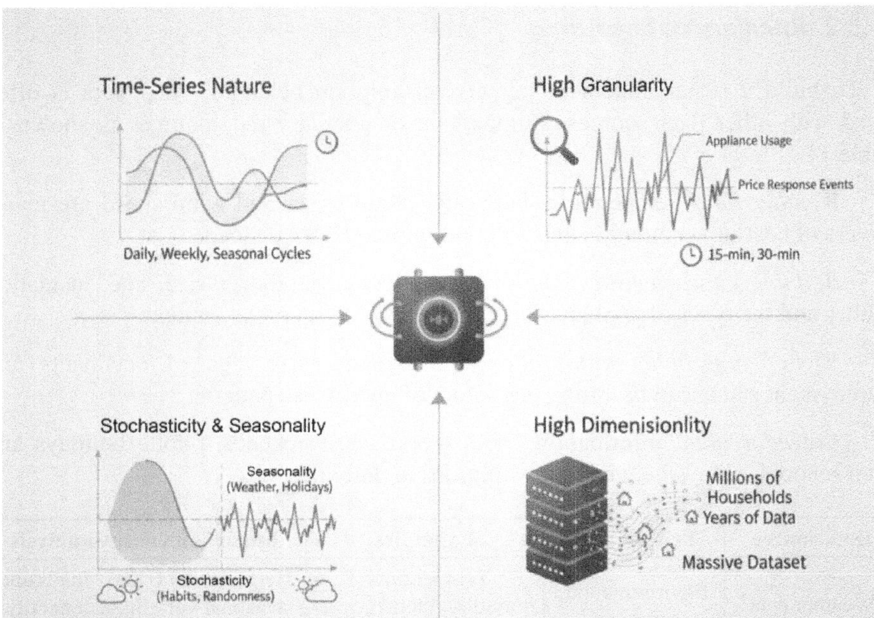

Figure 11.2. *Characteristics of smart meter data. For a color version of this figure, see www.iste.co.uk/kumar/windpower2.zip*

Smart meters record electricity consumption at high temporal resolutions, typically every 15, 30 or 60 minutes, a significant leap from the monthly readings of traditional meters. This data has several key characteristics:

– *Time-series nature*: the data is a sequence of measurements ordered chronologically. This temporal ordering is critical, as consumption patterns are deeply rooted in daily, weekly and seasonal cycles.

– *High granularity*: high-resolution data allows for the detailed analysis of consumption patterns, revealing events such as appliance usage, occupancy changes and responses to price signals.

– *Stochasticity and seasonality*: consumption is influenced by a multitude of factors, including weather, holidays and individual habits, leading to both predictable cycles (seasonality) and inherent randomness (stochasticity).

– *High dimensionality*: when considering thousands or millions of households over several years, the resulting dataset is massive.

11.3.2. *Ancillary data sources*

To build a richer understanding of consumption behavior, AMI data is often fused with other data sources. Comparison of ancillary data sources is shown in Table 11.1:

– *Weather data*: temperature, humidity, cloud cover and wind speed are major drivers of heating, ventilation and air conditioning (HVAC) loads.

– *Building characteristics*: information about a building's size, age, insulation quality and the types of appliances it contains provides crucial context.

– *Socio-demographic data*: household income, size, age of occupants and employment status can be strong predictors of energy use patterns.

– *Calendar data*: information about weekdays, weekends, public holidays and even school holidays help explain variations in daily rhythms.

Data source	Type of data	Examples	Impact on energy analysis
Weather data	Environmental, temporal	Temperature, humidity, cloud cover, wind speed	Helps model HVAC loads and seasonal variations in energy consumption
Building characteristics	Structural, technical	Building size, age, insulation, appliance types	Provides context for energy efficiency, appliance usage patterns and load estimation
Socio-demographic data	Human, household	Income, household size, occupant age, employment	Predicts energy consumption behavior and variability based on lifestyle and occupancy
Calendar data	Temporal/ event-based	Weekdays, weekends, public holidays, school holidays	Explains daily and weekly variations, peaks and troughs in energy usage

Table 11.1. *Comparison of ancillary data sources*

11.3.3. *Data challenges*

Working with real-world energy data presents several challenges that must be addressed before model training:

– *Data privacy*: smart meter data is highly sensitive. It can reveal information about a household's occupancy, daily routines and even the types of appliances being used. This makes data sharing for research and model training extremely difficult.

– *Missing values*: communication failures, meter malfunctions and data transmission errors can lead to gaps in the time-series data.

– *Data quality and noise*: meters can be inaccurate, and consumption data can contain anomalous readings or noise that needs to be filtered.

– *Data scarcity*: while AMI deployment is growing, high-quality, long-term datasets, especially those linked with rich metadata, can still be scarce, particularly for specific demographics or for testing new DR programs.

11.4. Applications of generative AI in energy consumption analysis

The ability to learn and replicate complex data distributions makes generative AI uniquely suited for several critical tasks in the energy sector.

11.4.1. *Synthetic data generation for privacy and augmentation*

This is arguably the most mature and impactful application of generative AI in this domain. By training a generative model (typically a GAN or VAE) on a real-world smart meter dataset, we can create an entirely new, synthetic dataset that shares the same statistical properties as the original but contains no one-to-one mapping to any real customer (Tascikaraoglu et al. 2019).

11.4.1.1. *Key benefits*

– *Privacy preservation*: synthetic data can be shared freely with researchers, startups and other stakeholders without compromising customer privacy. This overcomes a major bottleneck in energy innovation. The generated data is anonymous by design, as it does not correspond to any real individual.

– *Data augmentation*: for many machine learning tasks, such as load forecasting or non-intrusive load monitoring (NILM), the amount of available training data is a limiting factor. This is especially true for rare events, such as the consumption signature of a specific faulty appliance or the behavior during an extreme weather event. Generative models can be used to augment existing datasets, creating more varied and robust training sets and improving the performance of downstream models.

– *Fairness and bias mitigation*: if a dataset underrepresents a particular demographic group, models trained on it may perform poorly for that group. Generative AI can be used to rebalance datasets by generating more data for underrepresented classes, helping to mitigate algorithmic bias.

Specialized GAN architectures such as TimeGAN have been developed specifically for time-series data, ensuring that the generated load profiles preserve not just the point-in-time statistical distributions but also the complex temporal dynamics and autocorrelation inherent in energy data.

Example: an energy utility wants to run a data science competition to develop better forecasting algorithms. Instead of going through a lengthy and complex process of anonymizing real customer data, they use a GAN to generate a high-fidelity synthetic dataset of one million residential load profiles. This allows participants to innovate without any privacy risks.

11.4.2. *Probabilistic load forecasting and scenario generation*

Traditional forecasting models typically provide a point forecast a single best guess for future consumption. While useful, this approach fails to capture the inherent uncertainty of energy demand. Grid operators, however, need to understand the full range of possibilities to manage risk effectively.

Generative models, particularly those based on VAEs and transformers, excel at probabilistic forecasting. Instead of a single output, they can generate a large ensemble of plausible future scenarios.

– *Process*: a model is trained on historical load and weather data. To forecast, the current state (e.g. the last 24 hours of consumption) and a forecast for future weather are given. By sampling from the model's learned latent space multiple times, we can generate hundreds or thousands of different, but equally likely, future load profiles.

– *Benefit*: this provides a rich, data-driven probability distribution for future demand. A grid operator can then ask more sophisticated questions, such as: "What is the 95th percentile of demand tomorrow at 6 PM if a heatwave occurs?" or "What is the probability of an evening ramp rate exceeding 500 MW?" This allows for more robust planning, optimal scheduling of reserves and better risk management.

This is a significant step beyond simple confidence intervals, as the generated scenarios can capture complex, non-Gaussian dependencies and represent realistic system trajectories.

11.4.3. *Behavioral clustering and anomaly detection*

Understanding that not all consumers are alike is fundamental to designing effective energy efficiency and DR programs. Generative models offer a sophisticated approach to unsupervised customer segmentation. Applications of generative AI are shown in Figure 11.3.

– *Clustering*: a VAE can be trained on a large dataset of customer profiles. The encoder learns to map each profile to a point in the low-dimensional latent space. Because the VAE is encouraged to create a well-organized latent space, customers with similar consumption patterns will be mapped to nearby points. We can then apply simple clustering algorithms (such as K-Means) in this meaningful latent space to identify distinct behavioral groups (e.g. "Daytime Workers," "Night Owls," "High Weekend Users"). These data-driven segments are often more insightful than traditional segments based on simple metrics such as total consumption (Pan et al. 2022).

– *Anomaly detection*: a generative model trained on "normal" consumption data learns the distribution of typical behavior. When a new, unseen load profile is presented, the model can calculate the probability of that profile occurring under its learned distribution. Profiles with very low probability are flagged as anomalies. This could represent a faulty meter, electricity theft, a sudden change in occupancy or the failure of a major appliance such as an HVAC system. This approach is powerful because it does not require pre-labeled examples of anomalies to work.

11.4.4. *Personalized energy recommendations and demand response*

Generative AI can be a powerful tool for customer engagement. By learning an individual household's consumption patterns, a generative model can create personalized "what-if" scenarios:

– *Counterfactual generation*: the model can answer questions such as: "What would my energy profile have looked like yesterday if it had been 5 degrees cooler?" or "Show me how my profile would change if I shifted my EV charging from 7 PM to 11 PM." The model does this by manipulating input conditions (temperature, charging schedules) and generating the most likely corresponding load profile.

– *Personalized recommendations*: instead of generic advice ("Use less energy"), a utility can provide highly specific and compelling feedback. For example, an app could show a customer their generated "optimal" profile next to their actual profile, highlighting specific periods where shifting load could lead to significant savings. This visual, personalized and data-driven feedback is far more likely to drive behavioral change.

– *Optimizing demand response*: for DR programs, generative models can simulate how different customer segments are likely to respond to various incentives or price signals, allowing utilities to design and target programs with much greater precision.

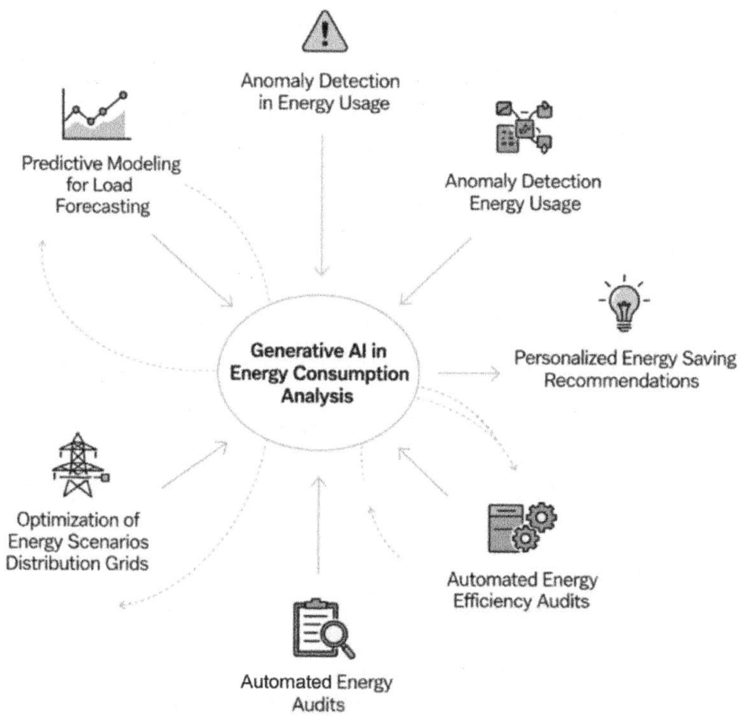

Figure 11.3. *Applications of generative AI. For a color version of this figure, see www.iste.co.uk/kumar/windpower2.zip*

11.5. Case studies

Energy utilities and policymakers increasingly face challenges in understanding and forecasting consumption patterns due to the variability of consumer behavior, data scarcity and the integration of renewable resources. Conventional statistical approaches often fail to capture the nonlinear and context-dependent nature of energy demand. Generative AI, with its ability to simulate diverse and realistic usage patterns, offers a transformative approach for tackling these challenges.

11.5.1. *Case study 1: synthetic household energy profiles using GANs*

A utility company piloted a project to generate synthetic household consumption datasets using cGANs. The GAN was trained on limited smart meter data (collected from ~5,000 households) and conditioned on external factors such as temperature, time-of-day and tariff type. Pictorial representation of this case study is shown in Figure 11.4:

– *Outcome*: the synthetic profiles closely resembled real-world consumption, preserving statistical properties such as peak demand and daily cycles.

– *Impact*: the enriched dataset improved the performance of forecasting models by 15% in mean absolute percentage error (MAPE) compared to models trained on original limited data. This also allowed the utility to run simulations of tariff changes without breaching customer privacy (Azzalini et al. 2025).

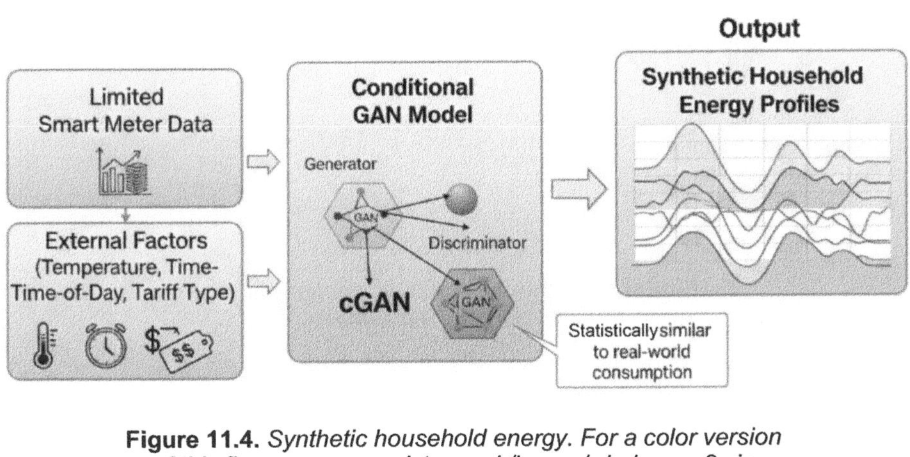

Figure 11.4. *Synthetic household energy. For a color version of this figure, see www.iste.co.uk/kumar/windpower2.zip*

11.5.2. *Case study 2: demand response optimization with VAEs*

A regional grid operator applied *conditional variational autoencoders (CVAEs)* to model behavioral responses to different DR signals. The CVAEs generated user-level scenarios under varying incentives, such as dynamic pricing and load-shifting rewards. Pictorial representation of this case study is shown in Figure 11.5:

– *Outcome*: the generated scenarios provided insights into how different consumer groups (residential vs. commercial) responded to DR signals.

– *Impact*: policymakers were able to design more effective DR programs, projecting a 12–18% reduction in peak load during high-demand hours.

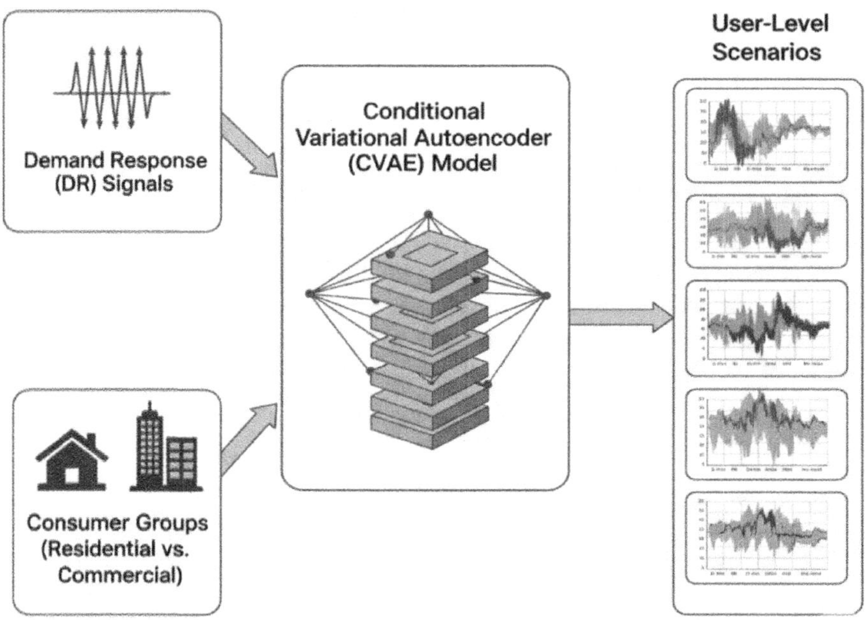

Figure 11.5. *Demand response optimization with conditional variational autoencoders*

11.5.3. *Case study 3: anomaly detection with flow-based models*

To enhance security, a national energy board deployed glow-based flow models to detect anomalous consumption patterns across industrial consumers. By learning probability distributions of normal usage, the system identified irregularities such as energy theft and equipment malfunctions. Pictorial representation of this case study is shown in Figure 11.6:

– *Outcome*: the model successfully flagged 92% of anomalous events, outperforming traditional rule-based systems.

– *Impact*: this resulted in early detection of theft worth millions in avoided losses annually and improved trust in automated monitoring systems.

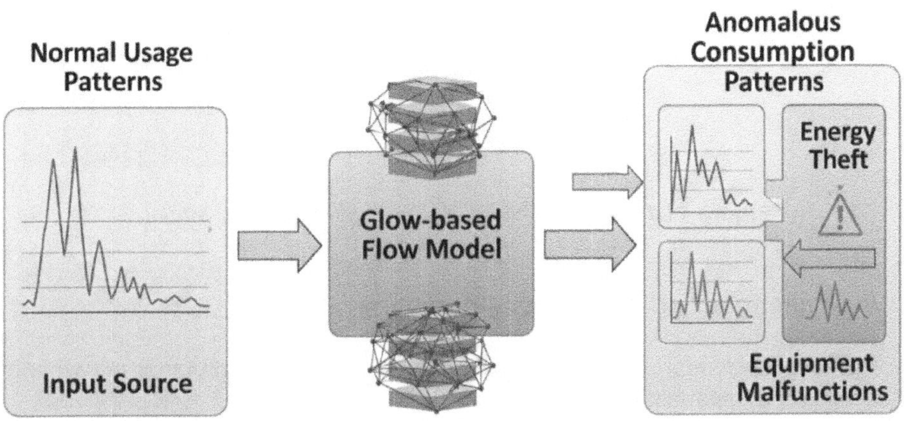

Figure 11.6. *Anomaly detection with glow-based flow models. For a color version of this figure, see www.iste.co.uk/kumar/windpower2.zip*

11.5.4. *Case study 4: behavior simulation with transformer models*

An energy research institute used transformer architectures to simulate long-term energy consumption under climate and policy scenarios (Mascali et al. 2023). By integrating weather forecasts, socio-economic indicators and appliance adoption rates, the model generated detailed household and regional consumption trends. Pictorial representation of this case study is shown in Figure 11.7:

– *Outcome*: the simulations highlighted how extreme weather events could increase residential cooling loads by 25–30% over the next decade.

– *Impact*: insights informed infrastructure planning, helping policymakers allocate renewable generation and storage resources more effectively.

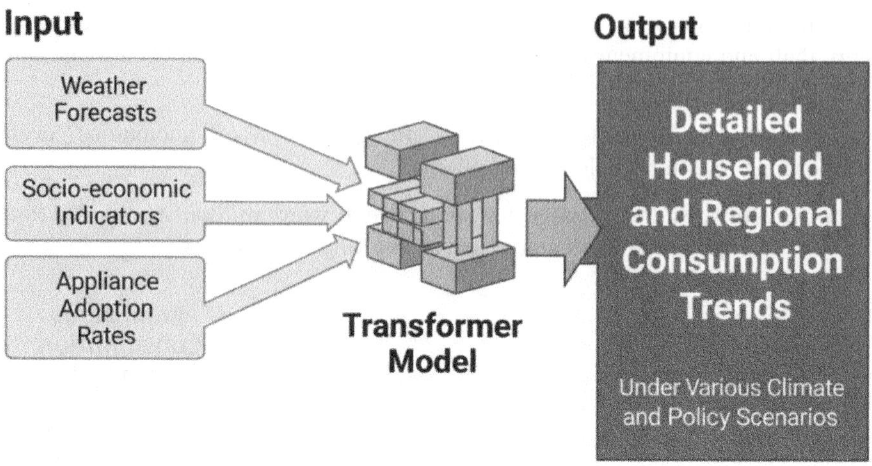

Figure 11.7. *Behavior simulation with a transformer model*

11.6. Challenges and ethical considerations

While the potential of generative AI is immense, its application is not without challenges and important considerations.

11.6.1. *Key challenges*

– *Computational cost*: training large, high-performing generative models such as GANs and transformers is computationally expensive, requiring significant GPU resources and time. This can be a barrier to entry for smaller organizations.

– *Model complexity and interpretability*: these models are often "black boxes". Understanding why a model generated a particular profile can be difficult, which can be a concern in highly regulated industries.

– *Evaluation difficulties*: as discussed, evaluating the quality of generated data is non-trivial. A lack of standardized, universally accepted metrics can make it hard to compare different models and approaches objectively.

– *Fidelity versus diversity*: there is often a trade-off between generating samples that are very close to the training data (high fidelity) and generating novel, diverse samples. Overfitting can lead to a model that simply memorizes and reproduces training examples.

11.6.2. *Ethical considerations*

– *Data privacy revisited*: while synthetic data enhances privacy, it is not a silver bullet. A poorly trained model could inadvertently leak information by generating samples that are too close to real individuals in the training set. Techniques from differential privacy can be integrated into the training process to provide mathematical guarantees of privacy, but this often comes at the cost of data utility.

– *Malicious use*: the ability to generate realistic data can be misused. For instance, synthetic load profiles could be used to design sophisticated attacks on the electricity grid by simulating false demand scenarios to manipulate energy markets.

– *Bias amplification*: if the original training data contains biases (e.g. it predominantly represents high-income households), the generative model will learn and potentially amplify these biases in the synthetic data it produces. This could lead to the development of energy services and technologies that are inequitable and serve some demographics better than others.

11.7. Conclusion

Generative AI represents a fundamental shift in our ability to analyze and understand energy consumption behavior. By moving beyond prediction and classification to the creation of new, realistic data, these models provide powerful tools to address long-standing challenges in the energy sector. From safeguarding customer privacy and enriching limited datasets to creating robust, probabilistic forecasts and enabling deeply personalized customer engagement, the applications are as diverse as they are impactful.

The path to widespread adoption requires overcoming significant technical challenges related to computational cost, model evaluation and interpretability, as well as navigating the critical ethical landscape of privacy and algorithmic bias.

However, the trajectory is clear. As generative models become more sophisticated, efficient and accessible, they will become an indispensable part of the analytical toolkit for utilities, grid operators, policymakers and researchers.

By learning the intricate language of our energy consumption, generative AI will not only help us understand the present but also empower us to actively generate a more efficient, resilient and sustainable energy future.

11.8. References

Asre, S. and Anwar, A. (2022). Synthetic energy data generation using time-variant generative adversarial network. *Electronics*, 11(3), 355. doi: 10.3390/electronics11030355.

Azzalini, D., Flammini, B., Emanuele, C.A., Guadagno, A., Ragaini, E., Amigoni, F. (2025). An empirical evaluation of deep autoencoders for anomaly detection in the electricity consumption of buildings. *Energy and Buildings*, 327, 115069. doi: 10.1016/j.enbuild.2025.115069.

Bohan, Z., Bin, Q., Sun, X., Li, W., Zhang, L., Guoqi, S., Wei, X., Qiong, W. (2023). DBSCAN-based electricity consumption anomaly detection method integrated with VAE. *Engineering Reports*, 5(12), e1263. doi: 10.1002/eng2.1263.

Bu, X., Wu, Q., Chen, J., Dong, J. (2021). A GAN-CNN based model for short-term load forecasting. *Energy Proceedings: Sustainable Energy Solutions for a Post-COVID Recovery*, 20, 46–52. doi: 10.46855/energy-proceedings-7925.

Dai, W., Liu, X., Heller, A., Nielsen, P.S. (2022). Smart meter data anomaly detection using variational recurrent autoencoders with attention. *Applied Energy*, 314, 118926. doi: 10.1016/j.apenergy.2022.118926.

Ji, T., Xu, P., Guo, D., Sun, L., Ma, K., Wang, Y., Han, X. (2025). Variational autoencoder-based anomaly detection in large-scale energy storage power stations. *Energies*, 18(11), 2770. doi: 10.3390/en18112770.

Lin, R., Chen, S., He, Z., Wu, B., Zou, H., Zhao, X., Li, Q. (2024). Electricity behaviour modeling and anomaly detection services based on a deep variational autoencoder network. *Energies*, 17(16), 3904. doi: 10.3390/en17163904.

Liu, X., Geng, J., Wang, Z., Cai, L., Yin, F. (2023). Data-driven energy efficiency evaluation and anomaly detection of multi-type enterprises based on big data mining. In *Big Data and Security – ICBDS 2022*, 1–18. Springer, Singapore. doi: 10.1007/978-981-99-6990-4_1.

Ma, Y.X. and Yeung, L.K. (2024). BEForeGAN: An image-based deep generative approach for day-ahead forecasting of building HVAC energy consumption. *Applied Energy*, 376, 121524. doi: 10.1016/j.apenergy.2024.121524.

Mascali, L., Schiera, D.S., Eiraudo, S., Barbierato, L., Giannantonio, R., Patti, E., Bottaccioli, L., Lanzini, A. (2023). A machine learning-based anomaly detection framework for building electricity consumption data. *Sustainable Energy, Grids and Networks*, 36, 101194. doi: 10.1016/j.segan.2023.101194.

Pan, H., Yin, Z., Jiang, X. (2022). High-dimensional energy consumption anomaly detection: A deep learning-based method for detecting anomalies. *Energies*, 15(17), 6139. doi: 10.3390/en15176139.

Razghandi, M., Zhou, H., Erol-Kantarci, M., Turgut, D. (2023). Smart home energy management: VAE-GAN synthetic dataset generator and Q-learning. *arXiv Preprint*, arXiv:2305.08885 [Online]. Available at: https://arxiv.org/abs/2305.08885.

Sánchez-Mompó, A., Mavromatis, I., Li, P., Katsaros, K., Khan, A. (2025). Green MLOps to green GenOps: An empirical study of energy consumption in discriminative and generative AI operations. *arXiv Preprint*, arXiv:2503.23934 [Online]. Available at: https://arxiv.org/abs/2503.23934.

Su, J., Chen, F., Qiu, Z., Zhong, Z., Zheng, H., Zheng, H. (2023). Multi-load short-term prediction of an integrated energy system based on GAN-LSTM. *Journal of Physics: Conference Series*, 2564(1), 012062. doi: 10.1088/1742-6596/2564/1/012062.

Sun, Y., Lee, J., Kim, S., Seon, J., Lee, S., Kyeong, C., Kim, J. (2023). Energy theft detection model based on VAE-GAN for imbalanced dataset. *Energies*, 16(3), 1109. doi: 10.3390/en16031109.

Tascikaraoglu, A., Erdinc, O., Uzunoglu, M., Karakas, A. (2019). Data-driven parallel prediction of building energy consumption using generative adversarial nets. *Energy and Buildings*, 186, 230–239. doi: 10.1016/j.enbuild.2019.01.013.

Yuan, R., Pourmousavi, S.A., Soong, W.L., Black, A.J., Liisberg, J.A.R., Lemos-Vinasco, J. (2022). A synthetic dataset of Danish Residential Electricity Prosumers. *Scientific Data*, 9(1), 288. doi: 10.1038/s41597-022-01378-3.

Zhang, N., Pan, X., Wang, Y., Zhang, M., Cheng, M., Shang, W. (2021). Adaptive IES load forecasting method based on the Octopus model. *Frontiers in Energy Research*, 9, 709708. doi: 10.3389/fenrg.2021.709708.

Zhang, W., Chen, C., Wang, J., Huang, T. (2023). A deep learning-based load forecasting algorithm for energy consumption monitoring system using dimension expansion. *EURASIP Journal on Advances in Signal Processing*, 2023(1), 102. doi: 10.1186/s13634-023-00996-w.

12

Wind Power Forecasting for Grid Stability Enhancement with Effective Integration of AI Techniques

The global shift towards sustainable energy has accelerated the deployment of wind power in modern grids. However, challenges such as energy intermittency, accurate power forecasting and grid stability remain significant barriers to large-scale adoption. This work explores the application of artificial intelligence (AI) algorithms to optimize the performance, reliability and stability of wind energy systems. Various AI techniques, including deep learning (DL), machine learning (ML), optimization methods and deep reinforcement learning (DRL), are used for critical tasks such as very short-term and short-term wind power forecasting, turbine fatigue load prediction, gearbox fault detection, adaptive power optimization and converter controller tuning in wind energy conversion systems to enhance grid stability and ensure seamless energy integration. By combining intelligent forecasting with grid management solutions, this study underscores the transformative potential of AI in stabilizing renewable energy systems and advancing sustainable power infrastructures. The results demonstrate that AI-driven wind power solutions improve forecasting accuracy, maximize power extraction, reduce mechanical stress on turbines and stabilize grid operations, contributing to resilient and sustainable energy infrastructures.

12.1. Introduction

The incorporation of renewable energy resources (RES) into electricity grids is necessary to improve the global energy system's efficiency and sustainability (Wang et al. 2024). Using RES such as solar, wind and hydroelectric electricity instead of conventional fossil fuels reduces reliance on finite resources, enhances energy

---

Chapter written by J. Johncy BAI, A. Lelin FRED, S. Jaisiva, V. VELMURUGAN and T. Dharma RAJ.

security and lowers greenhouse gas emissions. With the use of these abundant, eco-friendly resources, smart grids can diversify their energy mix, reduce the impacts of climate change and create a cleaner, greener future. But implementing RES into smart grids is fraught with several challenges that require innovative solutions to ensure their successful deployment (Xu et al. 2022). Since renewable energy is periodic and subject to elements such as weather and varying power production, sophisticated grid management systems must be used to ensure stability. A key component of these systems is accurate forecasting, which facilitates improved resource allocation and energy management decision-making. Furthermore, the decentralized nature of renewable energy sources makes it more challenging to optimize grid operations and manage power flows, underscoring the necessity for advanced forecasting frameworks to handle these problems successfully.

Wind power significantly contributes to balancing the generation and demand of electrical power. Scheduling conventional power units, as well as balancing the grid and managing reserve capacity, all rely on accurate forecasting (Zhao et al. 2023). To better accommodate wind energy converted into the electrical grid, continuous research and technology development focus on enhancing wind forecasting skills' precision and dependability. Global population expansion, urbanization, industrialization and technical improvements have all contributed to a steady increase in the demand for energy over time. Today's global environment has made ensuring that everyone has constant access to a plentiful supply of energy a top priority. The growing supply–demand imbalance and high electricity losses in developing countries such as India have made it more urgent for stakeholders to make strategic investments in the power sector. Several conventional (fossil fuels) and non-conventional (non-fossil fuels) energy sources can make electricity. All energy sources combined have an installed capacity of 8.9 TW, and India is ranked fourth in the world for both installed capacity and energy consumption. Figure 12.1 shows the wind energy.

Nuclear energy, coal, oil and natural gas are examples of conventional energy sources, and they are all finite. These sources have numerous disadvantages, including resource depletion, health effects, environmental harm, climate change and unstable economies, even though they are the primary sources of energy (Qiao et al. 2022). These are much less costly initially, but they have a great negative impact on the environment. Non-traditional, renewable and sustainable energy sources include hydropower, solar, wind, biomass, geothermal and tidal. These sources generate minimal carbon emissions and are environmentally benign. Since non-conventional energy requires new infrastructure and new technology, it is initially more expensive. With the advancement of technology, these expenses will soon drop and become a practical option.

Figure 12.1. *Wind energy. For a color version of this figure, see www.iste.co.uk/kumar/windpower2.zip*

The stability of the electrical supply in the future, transmission losses and pollution are only a few of the issues that society faces from conventional resource-based power generation. Given all these challenges, in order to fulfill present and future electrical demands, it is now essential to steadily increase the percentage of electricity generated from renewable sources. Among the various alternative energy sources, renewables are the best at producing power with the lowest pollution or carbon emissions in the process (Chen et al. 2021). Using these resources effectively in power generation can help mitigate air pollution, global warming and imbalances in energy supply and demand.

Renewable energy applications create cleaner, safer and more affordable energy in the future while also opening up new markets and professional opportunities. The need for economic expansion, environmental sustainability and energy independence highlights the significance that renewable energy for the global economy. Renewable energy can be used as a source of energy as well as a tool to fulfill several other important needs, such as energy security and reducing the adverse effects that greenhouse gas emissions and the use of fossil fuels have on human health and the environment.

This chapter examines how AI can improve wind power forecasting (WPF) and guarantee grid stability. It covers the basic elements of wind power generation, the methods and AI techniques used to make accurate forecasts, and methods for maintaining the electrical grid adaptable and stable in the face of renewable energy variability.

12.2. Overview of wind power prediction

WPF is one of the practices that involves the use of meteorological, turbine and grid data to forecast the quantity of energy that will be produced by wind systems in the future. Successful integration of renewable energy sources in power systems, optimization of the dispatch of energy and ensuring the stability of the grid and minimization of operational costs require accurate forecasting. Wind power is among the most promising sources of renewable energy that can be used to produce electricity. The process of using energy from wind and converting it to other useful forms of energy such as electricity is known as wind power and is made possible through the use of wind turbines. The production of wind energy has at times been criticized due to the fact that it is dependent on weather, making it unreliable and intermittent. The direct introduction of wind into a grid is impossible due to its unpredictable nature. Consequently, there is a dire need to ensure that correct predictions are made concerning the level of wind power to be produced. There should be a balance between electricity production and the quantity of electricity which is used to avoid electricity grid imbalances.

To make sure that the integration of wind power can be implemented with a reliable approach on a large scale, it is necessary to predict the wind power due to its unpredictable and intermittent nature. Various models may be applied to predict wind power. According to the prediction, dispatch, scheduling and unit commitments have to be arranged by the organization operators. The National Institute of Wind Energy (NIWE) is another independent research and development organization supervised by the Ministry of New and Renewable Energy (MNRE) which is a division of the government of India. It offers strong predictive services in generation scheduling in India. This helps ensure appropriate power management. NIWE created the Center of Innovation for Predicting Variable Wind and Solar Power. Tamil Nadu can get WPF services from NIWE. It is an independent research and development organization under MNRE. With a capacity of 7,400 MW and grid connection through substations, the forecasting services provided to Tamil Nadu are among the largest in the world. The range of the forecasting inaccuracy is between 2% and 6%.

In 2010, the WPF tool "Prediktor", created by the Danish Technical Institute Riso, was used by NIWE. This program predicts wind power using non-Gaussian forecasting and can be connected to any instrument or system used to collect wind data. In 2012, NIWE and Vortex, Spain, collaborated to enhance their ability to fulfill the demands of the Central Electricity Regulatory Commission (CERC) and Indian Electricity Grid Code (IEGC). They worked together on projects under the Hispano-Indian Bilateral Programme of Technological Cooperation (ISIP), in Renewable Energy. Among the other objectives was to provide services that would predict the wind power in Tamil Nadu. A combination of the physical and statistical models has been used to install NIWE/Vortex. NIWE is a state-based company that has been offering WPF services to the state of Tamil Nadu since June 2015. This has been achieved in collaboration with the Indian Wind Power Association. A project funded by the Indian Wind Power Association was geared towards supplying the infrastructure to all wind farm substations. The North India Wind Energy (NIWE) organization is applying the Vortex model to obtain the output of power generation of 107 substations of different wind farms distributed all over the state of Tamil Nadu. It is accessing this data via the General Packet Radio Service. The atmospheric data is collected and processed by the automated system. For 10 days ahead forecasting, the electricity generation is forecasted. The results of the prediction are updated every 1.5 hours for the next 10 days, according to the standards prescribed by the Central Electricity Regulatory Commission (CERC). The minimum accuracy for day-ahead scheduling should be 97%. However, for intraday scheduling, it should be 99%. NIWE provides forecasting data to Regional Loading Dispatching Centers and the Indian Wind Power Organization, and other stakeholders. There are numerous additional private companies that offer wind forecasting services.

12.2.1. *Wind power*

Wind power is an RES that generates power from the kinetic energy of moving air using wind turbines. Wind drives turbine blades to rotate, which generate electricity for a generator. It is a serious method of eliminating dependence on fossil fuels since it is renewable, clean and does not emit greenhouse gases in the process.

Wind energy can be harnessed in both onshore and offshore wind farms, and offshore wind farms have the advantage of having stronger and consistent wind. In order to be efficient, the contemporary wind farms include forecasting and smart grid technologies.

12.2.2. *Wind power generation components*

– Blades and rotor;

– pitch mechanism;

– low-speed shaft;

– gearbox;

– high-speed shaft and generator;

– controller;

– yaw drive and motor;

– tower and foundation.

12.2.2.1. *Blades and rotor*

The blades are aerodynamic devices made to transfer wind energy into rotational motion. For balance, noise reduction and maximum efficiency, the majority of contemporary turbines have three blades. The rotor spins as a result of lift being created on one side and drag being created on the other as wind passes over the blade surfaces (Liu et al. 2023). The rotor assembly joins the rotor to a central hub which delivers the torque to the low-speed shaft. As the power and wind speed cube are related, longer blades substantially increase the energy capture capacity. The blade's length directly impacts the swept area.

12.2.2.2. *Pitch mechanism*

The pitch mechanism modifies the blade's angle with respect to the airflow. This is necessary to maximize power output and control rotor speed. Blades are angled for optimal lift in moderate winds and away from the wind in strong gusts to minimize stress and avoid overspeed. There are two primary varieties: passive, centrifugal and aerodynamic-based, and active, operated by motors or hydraulics to adjust the blades individually. Additionally, pitch systems enable braking for maintenance or emergencies. In a variety of wind conditions, efficient pitch control increases operating safety, turbine longevity and energy efficiency.

12.2.2.3. *Low-speed shaft*

The rotor hub and the gearbox are attached to the low-speed shaft. The tremendous torque produced by the wind is transferred via the blades as it revolves at a speed of roughly 10–20 revolutions per minute (rpm). To withstand torsional loads and mechanical strains, this shaft needs to be strong. The shaft is supported by bearings, which lower friction and provide a smooth power transfer. A direct-drive turbine does not have a gearbox, so the low-speed shaft of a direct-drive turbine is directly connected to a generator. It is an essential component in the energy conversion chain because of its design, which takes fatigue resistance and vibration control into account to preserve efficiency and avoid mechanical breakdowns.

12.2.2.4. *Gearbox*

The generator is driven by the gearbox, which increases the rotational speed across its low-speed gear to the high-speed gear. It can, for instance, increase the generator's speed by 15 rpm to 1,500 rpm for best results. Bearings, lubrication systems and gear stages make up its smooth, effective operation. Although gears make compact generators possible, they need to be regularly maintained and are prone to wear. Systems for lubrication, cooling, lower friction and avoiding overheating are necessary. To save on maintenance, several contemporary turbines use direct-drive generators instead of gearboxes. But because they can effectively balance rotor speed with generator requirements, gearboxes are still widely used.

12.2.2.5. *High-speed shaft and generator*

The generator-high-speed shaft, which is attached to the gearbox, turns at a somewhat higher speed, typically 1,500 rpm. The generator transforms the mechanical energy when it is changed to electrical energy by electromagnetic induction. Synchronous generators and asynchronous (induction) generators are common types. Although wind speeds can vary, power converters can be used to maintain a constant output frequency. During operation, cooling systems such as liquid circuits or fans keep things from overheating. This is a crucial step in the wind energy conversion process since it is the moment at which mechanical energy is formally transformed into electrical power that can be used.

12.2.2.6. *Controller*

The controller controls the wind turbine's performance by monitoring variables such as temperature, vibration, power output, wind speed and rotor speed. The minimum and maximum wind speeds are used to determine whether to start or stop the turbine. Additionally, it regulates brake systems, yaw position and blade pitch to maximize efficiency and safeguard parts.

12.2.2.7. *Yaw drive and motor*

For optimal energy capture, the rotor is oriented towards the wind due to the yaw mechanism. It has an electric motor-powered yaw drive (gear mechanism) and a yaw bearing for rotation. To move the nacelle, the yaw motor is guided by inputs from the controller and wind vane. The blades experience less mechanical stress and aerodynamic losses when the yaw alignment is precise. Although the nacelle can be rotated 360 degrees by the system, movement is typically restricted to minor adjustments. For the structure to last longer and retain optimal turbine output, effective yaw control is crucial.

12.2.2.8. *Tower and foundation*

The tower lifts the rotor and nacelle up to higher altitudes, where wind is more steady and effective. Although concrete and hybrid designs are also used, modern towers typically consist of tubular steel for strength and ease of assembly, and they range in height from 80 to more than 150 meters. The foundation distributes loads into the Earth and serves as the structure's anchor. In contrast to offshore designs, which call for monopiles, jackets or floating platforms, onshore turbines frequently use gravity-based or piled foundations. The foundation must support not only the weight of the turbine but also seismic and wind pressures, and tower height and stability directly affect energy generation (Chaudhary et al. 2020).

12.2.3. *Wind power generation dynamics*

Dynamics of wind power generation are the physical, mechanical, electrical and regulatory procedures that convert the kinetic energy of airflow into consistent, grid-compatible electrical power. When wind passes over the turbine blades, aerodynamic lift is created, which starts the process and spins the rotor (Sulaiman and Mustaffa 2024). A drive train transfers this mechanical rotation to a generator, where it is transformed into electrical energy. In direct-drive systems, this transmission occurs directly, whereas in geared designs, it passes through a gearbox. The rotor inertia, aerodynamic properties, changeable wind conditions and the control systems used to maximize performance all affect a wind turbine's dynamic behavior. The system regulates rotor speed whenever the wind velocity decreases in order to ensure the optimal tip-speed ratio to achieve energy optimality. This mode is called maximum power point tracking (MPPT). In order to reduce power generation and to safeguard the turbine against mechanical overloading, the pitch control is enabled at higher wind velocities.

Variable-frequency output is converted to fixed-frequency AC to be connected to the grid, and generator types such as doubly fed induction generators (DFIG) or permanent magnet synchronous generators (PMSG) control electrical dynamics with power electronics. In addition to providing grid support services including reactive power control, voltage regulation and synthetic inertia, these power converters also use sophisticated control algorithms.

Equation [12.1] gives the following formula to calculate wind power as a result of wind speed, indicating how well the turbine converts wind energy from kinetic to electrical form:

$$P = \frac{1}{2} * A * \rho * V^3 * C_P \qquad [12.1]$$

– P: power output;

– A: area where the turbine blades sweep;

– ρ: air density;

– V: speed of wind;

– C_P: power coefficient.

The equation describes the relationship between the speed of wind, the efficiency of a turbine and the power output. Wind power prediction is very dependent on wind speed which has an influence on the effectiveness and precision of prediction. Because wind velocity and power output within wind turbines have a cubic relationship, any inaccuracy in the prediction of wind speed may result in large inaccuracies in predicted wind power production:

– *Wind capture*: by design, the blades of a wind turbine absorb the kinetic energy of the wind by slowing it down. They are typically built in places with strong winds, such as broad plains, coastal locations or high elevations.

– *Blade rotation*: the turbine rotates when the wind passes through the turbine. The speed of the wind is related to the speed of the rotating blade. The turbines that have been developed in the present generation are constructed to begin producing electricity when wind speed is relatively low and to produce the same as the wind speed increases.

– *Conversion of energy*: the revolving blades rotate a shaft, which is then attached to a generator installed in the nacelle mounted close to the top of the tower of the turbine. The rotating shaft is then transforming the mechanical energy to electricity in the generator.

– *Transmission of electricity*: transmission of electricity is the process by which the wind turbine's electricity is sent to the transformer located at the turbine base through cables inside the turbine tower.

– *Grid integration*: in a wind farm or park, electricity generated by many wind turbines is often collected. After that, this power is mixed with electricity from other sources and injected into the electrical system before being supplied to customers.

To make wind power a part of the electricity system that is reliable and efficient, several problems need to be solved:

– *Variability and intermittency*: because wind direction and speed vary so much, wind power is naturally very erratic. This intermittency makes it difficult for grid operators to keep a consistent supply–demand balance.

– *Grid stability*: the problem of sudden changes in the speed of the wind can make the grid unstable unless the changes in the production of power under its influence are properly managed. To guarantee system dependability, grid stability techniques such as voltage management and frequency regulation become essential.

– *Forecasting accuracy*: for effective grid operations, reliable wind power prediction is required. For grid operators to plan and make adjustments by wind energy generation predictions, forecasting technology must evolve.

WPF can be useful in the design and construction of a wind power system, in scheduling the production of electrical power, and in dispatching to the primary grid in order to meet the demand for power. It also makes it easier for traders, market analysts, wind farms and business owners to operate the power market efficiently. To maximize their profit, the anticipated data helps them make well-informed decisions on whether to buy or sell renewable energy (Al-Ibraheemi et al. 2024).

The multiple blade type (three to five number blades) model is highly suitable for small wind power generation systems. The main issue with this system is that the wind speed or velocity varies over time, meaning that the fluctuating electric power that is produced does not have a fixed value and is instead variable. Therefore, instead of immediately supplying the load shown in Figure 12.2, wind energy should ideally be fed into batteries or another form of storage device that can supply the load with the right amount of power, such as an accumulator circuit. Power generation in wind power systems increases in direct proportion to wind speed. Wind speeds can become too high to safely generate power, such as in rainy and stormy conditions. This is an environmentally balanced, pure and pollution-free power producing system.

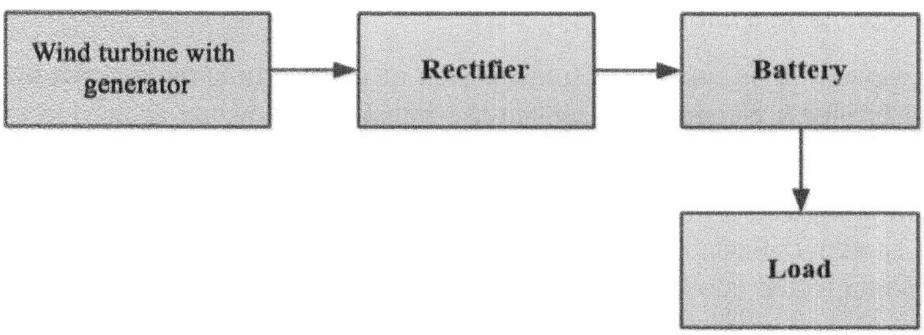

Figure 12.2. *Wind power system*

12.2.3.1. *A generic wind farm controller that maximizes power*

Upon providing a reference for power, the setup of a generic wind farm controller is shown in Figure 12.3, its purpose is to maximize the power output. Various inputs are fed to the controller which includes power produced, wind direction and velocity as well as the inclination angle. The efficiency of the wind turbines located behind the upstream turbines decreases as a result of the wakes. The major factor leading to the growth of auxiliary devices in wind farms is a decrease in power that is captured. Turbine failure is the ultimate result of non-rotating turbine sections becoming fatigued due to irregular wind loads. A reduction in power production and a rise in downstream turbine dynamic loading are two characteristics of wind wakes. The fluctuation in power generated by stochastic wind flow can pose a serious problem for wind farm operators.

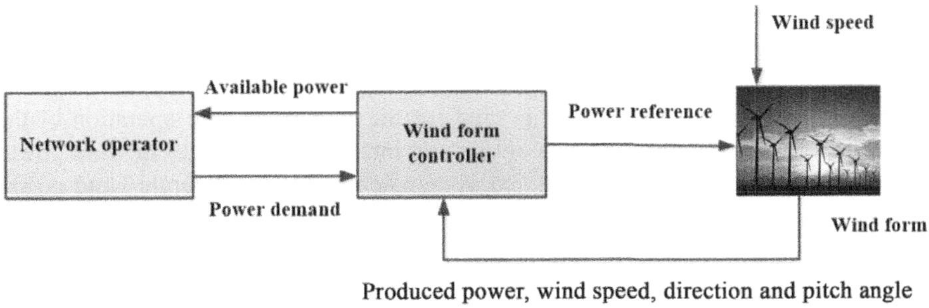

Figure 12.3. *A generic wind farm controller that maximizes power. For a color version of this figure, see www.iste.co.uk/kumar/windpower2.zip*

12.2.4. *Global wind power generation trends and perspectives*

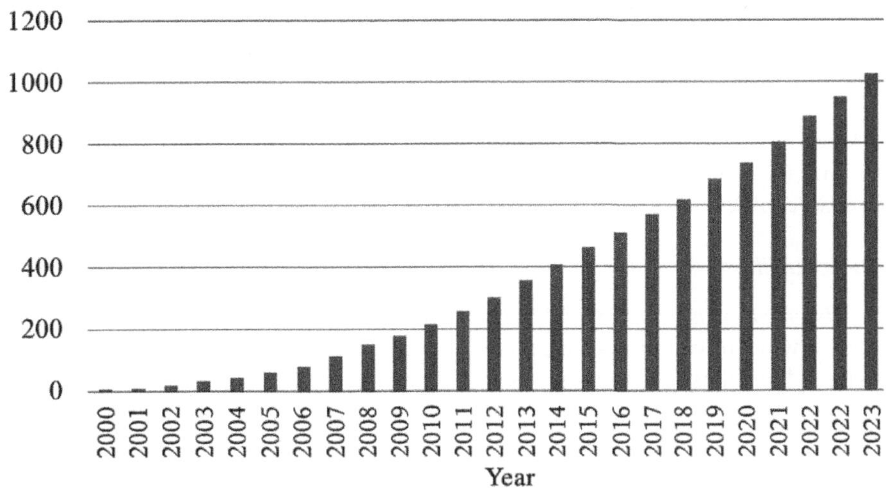

Figure 12.4. *Global wind power installed capacity (2000–2023)*

The majority of industry reports, databases and credible sources are used in the collection of global data regarding wind power installed capacity, these include national energy agencies, International Renewable Energy Agency (IRENA) and the Global Wind Energy Council (GWEC). According to the GWEC Global Wind Report, 2023 was the most successful year in the history of wind energy with a record of 117 GW of new capacity installed globally. The number of wind power systems installed globally is indicated in Figure 12.4.

12.2.5. *Importance and challenges of forecasting wind power*

Precise wind power forecasting is vital for the stable and safe operation of the power grid. Because wind energy is erratic and intermittent, changes in wind power have a big effect on load balancing and system scheduling. Inaccurate wind power forecasts can result in increased demand for backup power, frequency variations in the power system and potentially grid instability. Different time scales are used for WPF, which is often divided into short-, ultra- and long-term predictions, as well as scheduling plans, market trading and real-time control. However, low prediction accuracy and limited data are problems for wind power prediction because of the complexity of variables such as wind speed and weather.

12.3. Workflow of wind power prediction

The art of predicting the potential worth of wind energy which is generated by wind turbines in a certain location during a given time is known as forecasting wind power.

Depending on the needs of their operations, decision-makers also choose numerous look-ahead times for wind power predictions (between 14 minutes and days) (Gu et al. 2023). Figure 12.5 shows a workflow for forecasting wind power.

Data acquisition

– *Historical wind data*: wind farm's weather stations or anemometers record historical wind direction and speed information. In the case of the statistical and AI-based algorithms, it is the primary input information.

– *Meteorological data*: temperature, humidity and air pressure data are collected in real-time through satellites, weather stations or with the help of numerical weather prediction (NWP) models. In addition to historical data, these parameters are used to improve forecast accuracy.

Data pre-processing

Pre-processing of data entails preparing and cleaning the data collected to enable the data to be inputted into the forecasting model. The process of detecting and fixing errors, anomalies and inaccuracies in order to guarantee the quality of the collected information can be called data cleaning in WPF.

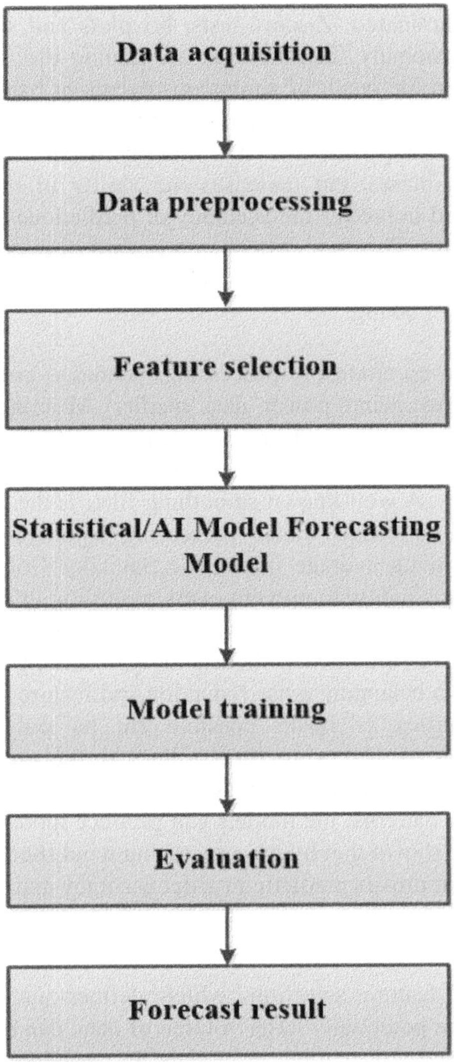

Figure 12.5. *Workflow of wind power forecasting*

Data pre-processing is a necessary process in order to make the wind power prediction models accurate and reliable. Extreme weather, communication failures and sensor malfunctions frequently affect wind power data, resulting in noise, outliers and missing values that can impair model effectiveness. Interpolation methods, such as linear or spline interpolation, use surrounding data to estimate gaps and correct missing values, even while large overall unrepresentative missing data may be completely eliminated. Z-score tests, boxplots and sophisticated machine learning (ML)-based anomaly identification are among the methods used to find outliers which are often the result of anomalous events or hardware failures. These methods guarantee the consistency and integrity of the dataset by locating and eliminating aberrant data points that substantially depart from typical patterns. Pre-processing lowers biases and increases the ability of the model to identify meaningful patterns and improves the accuracy of predictions through efficient data preparation and cleaning. This is a critical step toward developing the correct wind power forecast models to maximize grid integration, operational efficiency and aid energy market decision-making.

Digital filters are a common pre-processing technique used to minimize noise disturbance and enhance wind power data quality. Meaningful patterns can be obscured and prediction model performance deteriorated by noise in wind power data, which is frequently brought on by climatic variations, inaccurate sensors or communication failures. A well-known smoothing filter is the Savitzky–Golay filter that has been shown to reduce noise preserving significant local information. Contrary to simple moving average filters, the Savitzky–Golay filter incorporates equations into a sliding window to prevent oversimplifying information. substituting the value from the polynomial for each central point. Accurate wind power forecasts depend on the data being smoothed without unduly obscuring significant peaks and troughs. Its flexibility in balancing noise reduction and feature preservation based on particular data properties is made possible by its polynomial degree and configurable window size. By using these filters at the pre-processing stage to produce cleaner and more reliable datasets and using predictive models that concentrate on the key patterns, the models can produce more accurate and reliable forecasts. Studies have shown that by pre-processing wind speed and power data, the Savitzky–Golay filter improves prediction model accuracy and stability.

Feature selection

The procedure of feature selection, which defines and preserves the most relevant data among the potentially large volume of data, can be listed as enhancing the effectiveness of wind power forecast models. In the wind power information, a host of input parameters are often incorporated, such as temperature, pressure, wind direction, wind speed as well as knowledge about the turbine in operation. Although the characteristics can be used in forecasting, the addition of unnecessary or redundant features can make the model more difficult to predict, overfit and lead to a lower level of accuracy being attained. The feature selection task can be used to motivate such tasks to simplify the model and select the most important features. Consequently, the model is not as complex to compute and comprehend. Two common methods are recursive feature elimination (RFE), a process that removes features that are less important in succession with the aim of improving the performance of a model, and variable importance-based selection, such as Random Forest (RF) which ranks features by their contribution to the reduction of impurity. The feature selection process is a paramount element in the design of efficient, dependable and operational prediction models since it causes the most imperative predictive variables to become the center of focus, enhances the overall generalization of the prediction model, and offers an understanding of the factors that influence most of the production of wind power. Such techniques are applicable in establishing the main factors influencing wind power, which are temperature, air pressure, wind direction and wind speed. Time series features must be extracted and analyzed prior to modeling since wind power data shows notable periodic and temporal dependencies. To provide the model with greater temporal features, periodic components can be extracted from the data using the Fourier or wavelet transforms. To improve the model's time series learning capabilities, lag embedding is also frequently used to extract the impact of past data on present predictions.

Statistical/AI model forecasting model

Numerous methods are used in WPF, and the forecast horizon and data availability play a significant role in selecting the model. For very short-term forecasts, statistical techniques such as regression models and autoregressive integrated moving averages (ARIMA) work well because they draw on past trends and perseverance (Li et al. 2022a). The models of NWP are suitable in short- to medium-term predictions as they predict using atmospheric simulations to generate physics-based forecasts. The support vector machines (SVM), RF and gradient boosting are ML methods that are capable of modeling nonlinear weather-to-power output relationships. Examples of multi-layer deep learing (DL) neural networks (NNs) include the convolutional neural network (CNN) and long short-term memory (LSTM) networks, which are useful both to simulate complex temporal and spatial relationships, as well as to enhance the precision of short- and medium-term predictions.

Hybrid approaches combine many methods, such as ML or DL post-processing with NWP outputs, to capitalize on each method's advantages. Achieving a balance between forecast accuracy, computational cost and adaptability to shifting wind and grid conditions is necessary when choosing a model.

Model training

The data to be used in training a WPF model is often split into two main subsets, i.e. the training and the testing sets. The training set that usually includes 70–80% of the data is needed to fit the model, enables the model to find correlations, patterns and dependencies between the desired output and the input variables. In order to evaluate the performance of the model with unknown data, the remaining 20–30% is called the testing set. This division guarantees an objective evaluation of forecasting accuracy and helps identify overfitting.

Evaluation

To assess the performance of the trained model, test data is used. To ascertain how well the model extrapolates to new data, this step is crucial. To measure the accuracy and dependability of the model, several performance metrics are determined.

12.3.1. *Methods for forecasting wind power*

There are three processes in WPF: first, using a model to determine wind speed; second, estimating the forecast or projection of wind power output; and third, regional forecasting, also known as upscaling and downscaling are applicable across many time spans. A WPF system uses a wide range of sources of information such as NWP models, Supervisory Control and Data Acquisition (SCADA) data and local weather observations in which real-time information about the health of the wind power plants is included (Ma et al. 2022) in the list. Wind power prediction can be characterized with many characteristics, such as input data, time horizon, methodology and forecast output. Figure 12.6 presents the prediction methods of wind power.

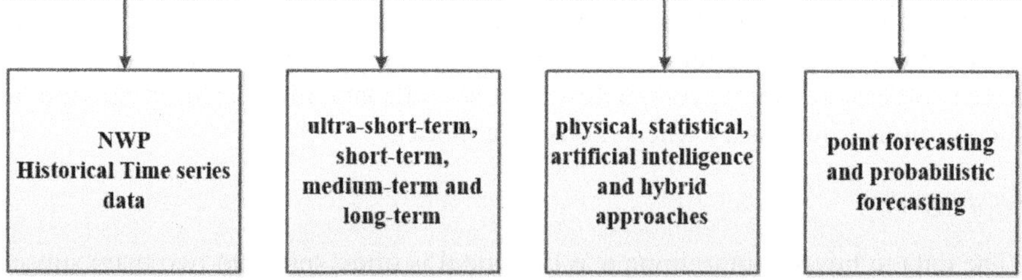

Figure 12.6. *Forecasting techniques for wind power*

12.3.1.1. *Input data*

Numerical Weather Prediction (NWP): NWP is a key component of WPF since it offers essential input data for modeling wind behavior. NWP uses computational algorithms and mathematical models to simulate atmospheric processes in order to produce weather forecasts. These predictions include factors such as air pressure, temperature, humidity and wind direction, and a variety of altitudes and spatial resolutions.

Time series data: wind power prediction is based on the historical time series data of wind turbine or wind farm power production, including measurements of wind direction and speed, to determine the forecaster's predictability, along with the patterns displayed by wind with respect to time. This time series data is available for a defined period (months or years) for the forecaster to develop and validate forecasting models. This data highlights the predictability of wind patterns to project future wind power generation.

12.3.1.2. *Time horizon*

WPF uses a variety of approaches and differs according to the time horizon; therefore, it is long-term, medium-term, short-term and ultra-short-term (Dong et al. 2022). For different stakeholders, this classification makes it easier to implement customized planning techniques and decision-making procedures. Table 12.1 shows the categories of WPF time horizons.

Forecast category	Time horizon	Typical application and advantage
Ultra-short-term	Seconds to one hour ahead	Supports grid operations, rapid regulatory actions and frequency regulation
Short-term	Half an hour to 24 hours ahead	Enables efficient load dispatch planning and short-range grid balancing.
Medium-term	One day to one week ahead	Guides unit commitment decisions, maintenance scheduling and operational planning.
Long-term	One week to several months ahead	Assists in production scheduling, grid integration strategies and long-term energy transition planning.

Table 12.1. *Categories of wind power forecasting time horizons*

WPF uses a variety of techniques, each drawing on different sources of information and using other types of computations and models, to predict wind farm power generation over time horizons of various lengths. The required accuracy, data accessibility and forecast duration influence the approach choice. Generally speaking, there are three categories of WPF approaches: AI, statistical and physical. Hybrid models sometimes use elements from other categories to achieve superior outcomes.

12.3.1.2.1. Short-term prediction

Short-term forecasting allows for efficient load dispatch planning and short-range grid balancing by projecting energy output and demand from 30 minutes to a full day in advance. This forecast horizon helps with day-to-day operational choices such as scheduling generating units, controlling reserves and maximizing the proportion of conventional and RES (Pei et al. 2022). Because RES are unpredictable and demand peaks are predicted, it is crucial for cost reduction. It is used to allocate storage resources, modify market bids and fine-tune generator set-points by utilities and grid operators. Precise short-term forecasting raises the dependability of renewable energy integration, lowers operational costs and increases efficiency.

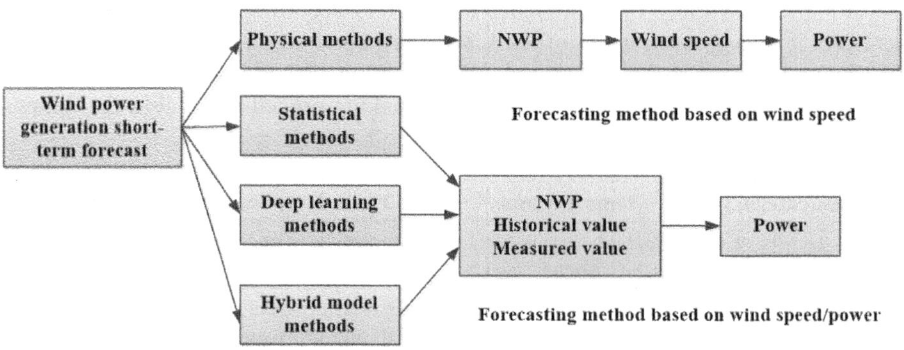

Figure 12.7. *Short-term forecasting of wind power generation*

According to the type of modeling technique, it is possible to classify four groups for forecasting wind power generation: DL, AI, statistical and physical techniques. Based on the different prediction parameters, it may be separated into two groups, as shown in Figure 12.7: direct corresponding power prediction and indirect corresponding wind speed prediction. Direct and indirect forecasting are the two categories into which WPF falls, per the forecasting model. The development of a comparable mathematical model based on the wind farm's previous wind power time series in order to forecast future wind power is known as direct prediction.

12.3.1.2.2. Medium-term forecasting

Guiding unit commitment, maintenance scheduling and overall operational planning, medium-term forecasting encompasses projections ranging from one day to one week ahead. Scheduling power plants, maximizing fuel purchases and planning grid maintenance without sacrificing stability all depend on it. Utility companies can better plan for future weather patterns that will impact renewable output and coordinate reserve requirements by predicting generating levels. This period is crucial for integrating solar and wind because it enables operators to match demand cycles with renewable forecasts. In the electrical markets, it also facilitates market trading methods, allowing participants to place well-informed bids in the days and weeks ahead.

12.3.1.2.3. Long-term prediction

Long-term prediction helps with capacity planning, production scheduling and long-term energy transition strategies by predicting energy patterns over weeks to months. For long periods, it allows utilities and regulators to plan for periods of high demand, optimize storage usage and create seasonal energy management strategies. Planning for the integration of renewable energy requires this prediction horizon to ensure that the infrastructure is prepared for anticipated seasonal fluctuations. It also influences financial hedging plans, contracts for the purchase of fuel and investment choices in new generation assets. Long-term projections help with climate resilience planning, renewable policy development and strategic grid extension, guaranteeing that energy systems continue to be economically and environmentally viable long-term.

12.3.1.2.4. Ultra-short-term prediction

Most often used for real-time grid operations, ultra-short-term forecasting offers quick projections of power generation in the range of seconds to an hour (Li et al. 2023b). It enables system operators to respond quickly to regulatory actions, maintain frequency stability and handle abrupt generation swings. It reduces the possibility of supply–demand imbalances brought on by abrupt changes in demand, wind or solar patterns by providing exact short-span forecasts. Automatic generation control (AGC), spinning reserve deployment and fast-response energy storage operation all depend on this category. Because of its high temporal resolution, power outages and other system disruptions are less likely to occur during unforeseen situations, ensuring grid stability.

12.3.1.3. *Methodology*

Several methods are used in WPF, each of which makes use of distinct information sources and computational techniques to anticipate wind farm power output over a range of periods. The required accuracy, data availability and forecast duration all influence the methodology selection. In general, WPF approaches fall into one of three types: statistical, physical and AI. For better results, hybrid models frequently combine components from several categories (Al-qaness et al. 2022).

12.3.1.3.1. Physical method

Applying a physical approach (deterministic approach), often referred to as the deterministic method, is predicted on NWP or weather predict data, including temperature, pressure, surface roughness and objects in the lower atmosphere. A number of physical models were created to estimate wind speed and forecasted wind power by using meteorological data. Large computer systems are necessary for physical models to produce reliable findings, even if they frequently depend on global meteorological data databases or atmospheric mesoscale models. The physical method estimates wind power output by providing an extensive account of the lower atmosphere. The local nature of the physical topography is considered using computational fluid dynamics (CFD) rather than using power law. Model output statistics (MOS) are also popular tools to minimize systematic forecasting errors and consider the unknowns in the predicted power generation (Zhu et al. 2022). Forecasts are given at particular nodes on a grid that displays data for a given region. The necessary height and location of the turbine hub must be added to these calculations since certain nodes do not have wind farms.

The physical approaches of WPF rely on sophisticated expertise of turbine environment and atmospheric processes to determine the possible wind farm power generation. These techniques are used to replicate wind flows, and they do this by creating a model of the atmosphere; they consider the changes in temperature, the gradients in the changing pressures as well as the topographical characteristics. However, a number of studies have demonstrated that experimentally observed power lines have an edge over theoretical ones, hence diminishing the significance of theoretical power curves. The physical technique was used to permanently model the process that regulates wind generation from NWPs at different sites around a wind farm. After that, the transfer function is computed and applied to the current weather predictions. MOS are often used in physical simulations for post-processing power estimates to correct for systematic forecasting errors caused by the NWP model or modeling methodology. Statistical forecasting methods are based on one or more models that indicate the relationship between past values of power, current measurements of wind power, and future and past values of meteorological variables with each other.

12.3.1.3.2. Statistical methods

The statistical method ignores weather conditions and is based on a large amount of previous data. AI (neuron-fuzzy networks, NNs) and time series analysis techniques were typically used. A wind farm power curve model that is semi-parametric and accounts for wind direction and speed is part of the statistical model set, as are dynamic forecasting models that explain how wind power changes in response to weather variations, etc. Statistical forecasting methods are based on the assumption that one or more models exist that exhibit the relationship between historical values of power, wind power measurements, and previous and future values of meteorological variables. Even while problem expertise is needed to choose the right meteorological variables and create the right models, the physical occurrences are not dissected and taken into consideration. The model parameters are determined by using a set of already known data, which are then updated often while the system is operating online to take into consideration any new information that becomes available (such as power measurements and weather forecasts). There are structural and black-box statistical models, linear and nonlinear statistical models. While black-box models are constructed mechanically from data and require minimal subject–matter expertise, structural models depend on the analyst's understanding of the phenomenon of interest. Modeling daily variations in wind speed or particular weather-related functions are examples of architectural techniques in WPF. Two types of black-box models are NNs and SVMs. On the other hand, other models fall somewhere "in-between" being wholly structural and being wholly black box. Examples of expert systems include those that can be designed with previous knowledge and that learn from experience (from a dataset). Next up is the topic of gray-box modeling. Two components are frequently included in statistical models: an autoregressive component that captures the persistent nature of the wind within a "meteorological" component that transforms predictions of meteorological variables in a nonlinear way.

Currently, the primary focus of statistical techniques to wind power prediction is on warnings about potentially high uncertainty or the optimal utilization for geographically scattered data from measurements for prediction error correction. Combining several weather forecasts as the input is another area of focus. The statistical correlation between the production of wind energy and the necessary input data is determined. They involve directly translating the input elements into wind generation through the use of a statistical model. With the use of these models, wind power can be determined in a single step using input parameters. Most models based on time series analysis methods (ARIMA, fractional ARIMA, etc.) and data mining methods (SVM, fuzzy model, ANN, model trees, etc.) can be used as output models. The statistical method processes enormous amounts of data and therefore does not provide a clear representation of meteorological phenomena. This is used to predict future power output by determining the relationship between weather and previous power output. As compared to the physical methodology, the statistical methodology only necessitates transformation of input variables to energize output. In contrast to physical methods, statistical methods simply require a single transformation of input variables to power output. Consequently, the processes that are used are called "black box". Most of the time, a statistical correlation is found between the wind farm's potential power output and the weather forecast or projection. Mathematical techniques such as autoregressive (AR), Kalman filter, Box–Jenkins methodology, autoregressive moving average model (ARMA), moving average (MA) and autoregressive integrated moving average model (ARIMA) are also used. Statistical methods such as ARIMA, persistence and AI algorithms are commonly used to make short-term predictions (He et al. 2022). These models are based on the trends in historical data.

12.3.1.3.3. AI methods

AI applications in wind power prognostication encompass diverse advanced ML approaches that are intended to predict future wind power generation. These methods make use of past data of wind direction, speed and power production, and meteorology to train predictive models. ML algorithms rely on meteorological variables, historical wind data and other applicable variables to offer the true approximations of the wind power production. Typical ML-based methods for wind energy forecasting include the following: RF, Gaussian Process Regression (GPR), SVM and Gradient Boosting Machines (GBM), such as XGBoost and LightGBM, are used for probabilistic forecasting with quantification of uncertainty.

Recent architectures in DL include deep neural networks (DNNs) for capturing high-dimensional relationships, recurrent neural networks (RNNs) and LSTM networks for modeling temporal dependencies in time series, and gated recurrent units (GRUs) for efficient sequence learning. CNNs are increasingly being used to process spatiotemporal wind field data, which enables the extraction of spatial information from weather maps or satellite imagery. By combining LSTMs for temporal modeling with CNNs for spatial feature extraction, hybrid DL models have enhanced their capacity to detect both long-term trends and short-term variability. The most common problems being tackled by optimization techniques (such as Simulated Annealing (SA), Genetic Algorithms (GA) and Particle Swarm Optimization (PSO)) are feature selection, ensemble model weighting and hyperparameter tuning to minimize the error in prediction (Ahmadi et al. 2022). DRL, which combines DL and reinforcement learning, enables adaptive forecasting by leveraging reward-based feedback and allowing models to interact with dynamic environments. Though they need a lot of high-quality training data and careful tuning to avoid overfitting, the primary benefits of AI techniques are their adaptability to non-stationary settings, flexibility and capacity to incorporate disparate data sources.

12.3.1.4. *Forecasting results*

Wind power forecasting generates two types of outcomes, point forecasting and probabilistic forecasting. Each has specific uses and provides unique insights.

12.3.1.4.1. Point forecasting

The point or deterministic forecasting technique gives one precise prediction of the future output of wind power at a given time. There are many methods for creating point forecasts, including statistical models such as ARIMA and ML algorithms such as SVMs and ANNs. These methods make use of both historical wind data and the current weather conditions to calculate the most anticipated wind power consumption.

12.3.1.4.2. Probabilistic forecasting

Unlike a single deterministic forecast, probabilistic WPF provides a range of possible results for future wind power generation. The unpredictability of wind direction and speed has a direct impact on power generation. Probabilistic forecasting provides a more detailed picture of the potential wind power outputs, and it starts with the construction of quantiles, prediction intervals and probability distributions. This approach will enable more informed decisions in energy planning, grid operation and integration of renewable energy, and help the stakeholders to understand and manage the uncertainty surrounding the generation of wind power.

12.3.2. *Predicting wind power and spatial transferability at various locations*

12.3.2.1. *Models for predicting spatial transferable wind power*

To manage the operational variability of wind farms, probabilistic ML and geospatial features have been combined to produce spatially transferable models. Power output is accurately predicted by these models under a variety of geographical circumstances. Analysis shows that while drastically lowering computational overhead, ML models closely matched observed data. In geographically diversified wind energy systems, this method is especially useful for dynamic operations and real-time decision-making.

12.3.2.2. *Models for multi-location prediction and flexibility*

Combining CNNs with LSTMs allows for accurate wind power forecasts at several sites by using each system's advantages in temporal and spatial data analysis. These hybrid structures operate well in a variety of locations with varying topography and climate. These models are essential for worldwide renewable energy applications because of their versatility, which guarantees reliable performance. Additionally, this modeling approach supports sustainable energy solutions by improving grid integration by taking into account a variety of environmental factors.

12.3.3. *Features of an effective forecasting model*

For WPF models to be effective, they must have the following features:

Accuracy: this is the most crucial component of a successful forecasting model. A number of variables, such as the state of the data at hand, the different forecasting techniques, the projected time horizon, along with the specifics on the wind farm or region, can affect how accurate the forecast is. The inherent fluctuation of wind speeds and directions, in addition to the impact of meteorological conditions, can make it challenging to forecast wind power with high accuracy.

Generalizability: beyond its training dataset, the model's capacity to produce accurate predictions on unknown and unseen data is essential. To make that the algorithm does not overfit according to the set of training information and that it can generate precise forecasting results on unknown data, evaluation procedures can also be put in place.

Flexibility: the model should be flexible enough to change over time or adapt to the changes in patterns or distribution of its underlying data. This is crucial for forecasting models because many things can affect the weather.

Considering uncertainty: an efficient forecasting model must have uncertainty about its forecasts. The uncertainty may be in the form of probability distributions, prediction intervals or confidence intervals. Three measures of uncertainty exist which are prediction interval normalized average width (PINAW), coverage width criterion (CWC) and prediction interval coverage probability (PICP).

When creating a forecasting model, these characteristics are constantly considered.

12.4. Grid stability

The capacity of an electric power system to function within acceptable bounds while providing customers with consistent, high-quality electricity is known as grid stability. Essentially, it guarantees that generator synchronism, voltage and frequency are maintained even in the event of grid disruptions, ranging from moderate (such as steady load variation) to severe (such as instantaneous generator trip) (Ahmed et al. 2023). To avoid cascading failures that could result in widespread blackouts, a reliable grid is necessary. Coordination of generating units, transmission systems and load control techniques results in stability. The system's strength, operational reserves and the responsiveness of its control mechanisms all affect the stability margin. Because power systems are dynamic by nature, their response to disturbances depends on several physical factors, including electrical inertia, reactive power balance and control loop performance. The combination of distributed generation, flexible loads and renewable energy sources has complicated modern grids, making stability a more difficult task. The power system stability classification is displayed in Figure 12.8.

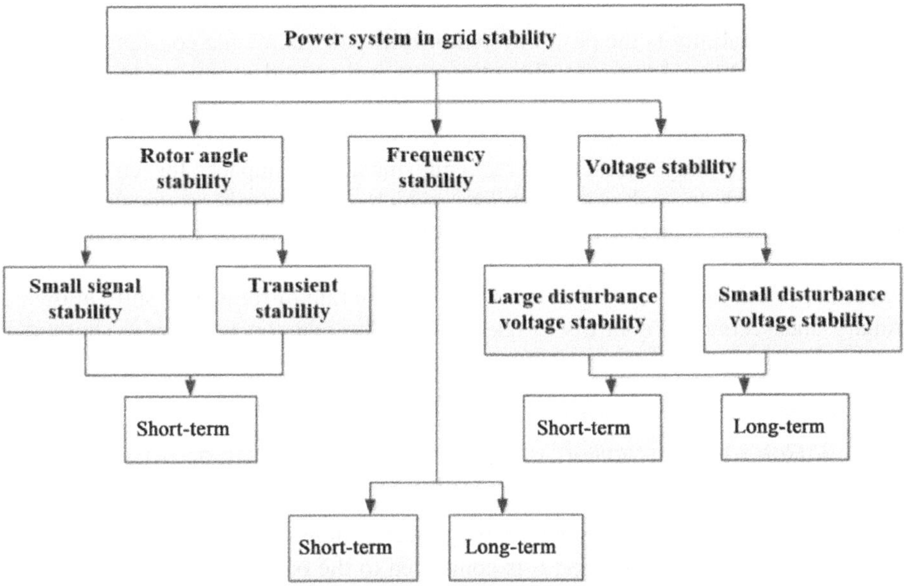

Figure 12.8. *Power system stability classification*

12.4.1. *Types of grid stability*

– Voltage stability;

– frequency stability;

– rotor angle stability (synchronous stability).

12.4.1.1. *Voltage stability*

Voltage stability is the capacity of the power system to maintain acceptable voltage at all the buses within the system, both under normal and stressed systems. Instability develops when the system is not able to supply the reactive power demanded by the loads and transmission network and this may lead to uncontrollably high voltages or smooth decreases of voltages. A gradual voltage drop that quickens when reactive support is insufficient frequently precedes voltage collapse. Long transmission distances, heavily loaded lines, inadequate reactive adjustment and abrupt, significant load increases are some of the causes. The issue is even worse in renewable-rich grids since solar and wind inverters may not be as capable of handling reactive electricity as synchronous equipment.

12.4.1.2. *Frequency stability*

Frequency stability is the power system's ability to maintain a constant operating rate in the face of a significant disruption that has caused a noticeable imbalance between generation and demand. The nominal frequency in the majority of countries is either 50 Hz or 60 Hz, and deviations beyond ±0.5 Hz may result in equipment faults or precautionary measures. Frequency instability impacts the entire linked grid, in contrast to voltage instability, which is primarily a local problem. Disturbances that can cause frequency deviation include sudden changes in load or transmission line issues or abrupt trips in the generators. Large synchronous generator system inertia has historically reduced the rate of frequency shift, allowing primary and secondary controls to take effect. The reaction window is shortened, though, by the increasing proportion of inverter-based renewable power, which changes frequency considerably more quickly due to its lack of natural inertia.

12.4.1.3. *Rotor angle stability*

The ability of the power system to continue synchronization with disruption by synchronous generator capacity within the power system is known as the rotor angle stability or synchronous stability. It is connected to the balance between mechanical and electromagnetic torques (produced by the prime mover and the generator, respectively). Rotor speed acceleration or deceleration may come from abrupt changes in electrical power transfer brought on by a disturbance, such as a transmission line short circuit. Large-scale blackouts and system separation may result from the rotors of the generators losing synchronization. There are two types of rotor angle stability:

– *Small-signal stability*: this refers to the system's capacity to remain synchronized in the face of minor disruptions, frequently caused by inadequate oscillation damping.

– *Transient stability*: this is the ability of the system to restart in synchronization within a few seconds following considerable disturbances, for example, failure or sudden load variation.

12.4.2. *Characteristics of grid stability*

– *Frequency*: variability in wind leads to variations in grid frequency. Fast-reacting reserves, governor controls and wind turbine-generated synthetic inertia are all necessary to maintain stability. In high wind penetration systems, frequency management is crucial because a steady frequency guards against equipment damage and guarantees constant power quality.

– *Voltage*: variations in wind output that occur quickly can affect grid-wide voltage levels. Reactive power compensation, FACTS devices and sophisticated turbine control systems all contribute to the maintenance of voltage stability. In changing generation conditions, stable voltage avoids blackouts caused by voltage breakdown and guarantees dependable electricity delivery.

– *Flexibility requirements*: demand response systems, battery storage and fast-ramping generating units are necessary for grid flexibility when integrating wind energy. By using these steps, the grid can adjust to abrupt variations in production and effectively strike a balance between supply and demand, particularly wind that fluctuates quickly.

– *Resilience to intermittency*: grid stability systems are built to endure the sporadic nature of wind. Despite the erratic availability of wind resources, systems maintain dependable electricity delivery through the deployment of energy storage, diverse generation portfolios and predictive operating tactics.

12.5. Conclusion

This chapter presents the revolutionary potential of AI for the renewable energy industry by providing a thorough examination of its applications in WPF and grid stability improvement. Forecasting accuracy is significantly increased by using cutting-edge AI techniques, including DL, ML, hybrid optimization and DRL. This makes it possible for dependable scheduling, effective load dispatch and proactive grid stabilization. Precise forecasts maintain stability in voltage, frequency and rotor angle throughout the power system, lessen the adverse effects of wind intermittency, save operating costs and increase turbine efficiency.

This chapter discusses some of the basic components and dynamics of wind power generation, the global trends and how the process of forecasting is done, including data collection, pre-processing, model selection and evaluation. The importance of optimization algorithms in enhancing prediction accuracy is examined together with a variety of approaches, including statistical, physical, AI-based and hybrid. Additionally, AI-driven methods allow for flexible reactions to quickly shifting meteorological circumstances, guaranteeing smooth integration of renewable energy sources. These technologies, when paired with intelligent grid control, are crucial to creating low-carbon, resilient and sustainable energy systems that can satisfy future demands while reducing their adverse effects on the environment.

12.6. References

Ahmadi, A., Talaei, M., Sadipour, M., Amani, A.M., Jalili, M. (2022). Deep federated learning-based privacy-preserving wind power forecasting. *IEEE Access*, 11, 39521–39530.

Ahmed, F., Al Kez, D., McLoone, S., Best, R.J., Cameron, C., Foley, A. (2023). Dynamic grid stability in low-carbon power systems with minimum inertia. *Renewable Energy*, 210, 486–506.

Al-Ibraheemi, Z. and Al-Janabi, S. (2024). Sustainable energy: Advancing wind power forecasting with grey wolf optimization and GRU models. *Results in Engineering*, 24, 102930.

Al-qaness, M.A., Ewees, A.A., Fan, H., Abualigah, L., Abd Elaziz, M. (2022). Boosted ANFIS model using augmented marine predator algorithm with mutation operators for wind power forecasting. *Applied Energy*, 314, 118851.

Chaudhary, A., Sharma, A., Kumar, A., Dikshit, K., Kumar, N. (2020). Short-term wind power forecasting using machine learning techniques. *Journal of Statistics and Management Systems*, 23(1), 145–156.

Chen, H., Birkelund, Y., Zhang, Q. (2021). Data-augmented sequential deep learning for wind power forecasting. *Energy Conversion and Management*, 248, 114790.

Dong, Y., Ma, S., Zhang, H., Yang, G. (2022). Wind power prediction based on multi-class autoregressive moving average model with logistic function. *Journal of Modern Power Systems and Clean Energy*, 10(5), 1184–1193.

Gu, B., Hu, H., Zhao, J., Zhang, H., Liu, X. (2023). Short-term wind power forecasting and uncertainty analysis based on FCM–WOA–ELM–GMM. *Energy Reports*, 9, 807–819.

He, B., Ye, L., Pei, M., Lu, P., Dai, B., Li, Z., Wang, K. (2022). A combined model for short-term wind power forecasting based on the analysis of numerical weather prediction data. *Energy Reports*, 8, 929–939.

Li, Z., Luo, X., Liu, M., Cao, X., Du, S., Sun, H. (2022). Wind power prediction based on EEMD-Tent-SSA-LS-SVM. *Energy Reports*, 8, 3234–3243.

Li, N., Dong, J., Liu, L., Li, H., Yan, J. (2023). A novel EMD and causal convolutional network integrated with Transformer for ultra short-term wind power forecasting. *International Journal of Electrical Power & Energy Systems*, 154, 109470.

Liu, L., Liu, J., Ye, Y., Liu, H., Chen, K., Li, D., Dong, X., Sun, M. (2023). Ultra-short-term wind power forecasting based on a deep Bayesian model with uncertainty. *Renewable Energy*, 205, 598–607.

Ma, Z. and Mei, G. (2022). A hybrid attention-based deep learning approach for wind power prediction. *Applied Energy*, 323, 119608.

Pei, M., Ye, L., Li, Y., Luo, Y., Song, X., Yu, Y., Zhao, Y. (2022). Short-term regional wind power forecasting based on spatial–temporal correlation and dynamic clustering model. *Energy Reports*, 8, 10786–10802.

Qiao, B., Liu, J., Wu, P., Teng, Y., (2022). Wind power forecasting based on variational mode decomposition and high-order fuzzy cognitive maps. *Applied Soft Computing*, 129, 109586.

Sulaiman, M.H. and Mustaffa, Z. (2024). Enhancing wind power forecasting accuracy with hybrid deep learning and teaching-learning-based optimization. *Cleaner Energy Systems*, 9, 100139.

Wang, H., Li, B., Xue, Z., Fan, S., Liu, X. (2024). Powerformer: A temporal-based transformer model for wind power forecasting. *Energy Reports*, 11, 736–744.

Xu, Y., Jia, L., Yang, W. (2022). Correlation-based neuro-fuzzy Wiener-type wind power forecasting model by using special separate signals. *Energy Conversion and Management*, 253, 115173.

Zhao, Z., Yun, S., Jia, L., Guo, J., Meng, Y., He, N., Li, X., Shi, J., Yang, L. (2023). Hybrid VMD-CNN-GRU-based model for short-term forecasting of wind power considering spatio-temporal features. *Engineering Applications of Artificial Intelligence*, 121, 105982.

Zhu, J., Su, L., Li, Y. (2022). Wind power forecasting based on new hybrid model with TCN residual modification. *Energy and AI*, 10, 100199.

List of Authors

R. ANUJA
Department of EEE
Arunachala College of Engineering for Women
Nagercoil
India

J. Johncy BAI
Department of EEE
Arunachala College of Engineering for Women
Nagercoil
India

E. Afreen BANU
Department of Computer Engineering
Shah and Anchor Kutchhi Engineering College
Mumbai
India

Malathy BATUMALAY
Centre for Data Science and Sustainable Technologies
INTI International University
Nilai
Malaysia

Suman CHAHAR
University Institute of Computing
Chandigarh University
Mohali
India

Ashutosh Kumar DUBEY
Department of Computer Science
School of Engineering and Technology
Chitkara University
India

Mary A.G. EZHIL
Department of EEE
Arunachala College of Engineering for Women
Nagercoil
India

A. Lelin FRED
Department of CSE
Mar Ephraem College of Engineering and Technology
Marthandam
India

R. GAYATHRI
Senior Engineer
TVS Brakes India Pvt. Ltd.
Chennai
India

S. JAISIVA
Department of EEE
Sri Krishna College of Engineering and Technology
Coimbatore
India

J.D. Dorathi JAYASEELI
Department of Computing Technologies
College of Engineering and Technology
SRM Institute of Science and Technology
Chennai
India

J. Reyes JUÁREZ-RAMÍREZ
Facultad de Ciencias Quíquimicas e Ingeniería
Universidad Autónoma de Baja California
Mexico

Abhishek KUMAR
Department of Computer Science and Engineering
Chandigarh University
India

D. LAKSHMI
Department of EEE
AMET (Deemed to be University)
Chennai
India

Mamta
Department of CSE
Chandigarh University
Mohali
India

Komal MISHRA
University Institute of Computing
Chandigarh University
Mohali
India

A.S. MONIKANDAN
Department of EEE
Arunachala College of Engineering for Women
Nagercoil
India

M. Dhiviya NYCIL
Department of EEE
Arunachala College of Engineering for Women
Nagercoil
India

Rajasekaran PALANIAPPAN
Department of Computing Technologies
College of Engineering and Technology
SRM Institute of Science and Technology
Chennai
India

T. Dharma RAJ
Department of EEE
Muthayammal Engineering College
Rasipuram
India

R. RAJASREE
Department of EEE
AMET (Deemed to be University)
Chennai
India

Mantena Siva Pavan Kumar RAJU
Center for Innovation and Entrepreneurship
Sasi Institute of Technology and Engineering
Tadepalligudem
India

Neetu RANI
Department of Computer Science and Engineering
Chandigarh University
Gharuan
India

P. ROBERT
Department of Computing Technologies
College of Engineering and Technology
SRM Institute of Science and Technology
Chennai
India

Mantena SIREESHA
Center for Geospatial and Saline Studies
Sasi Institute of Technology and Engineering
Tadepalligudem
India

Arun Lal SRIVASTAV
School of Engineering and Technology
Chitkara University
Himachal Pradesh
India

M. SUTHANTHIRA
Department of EEE
Arunachala College of Engineering for Women
Nagercoil
India

Ananth Kumar T.
Department of Computer Science and Engineering
IFET College of Engineering (Autonomous Institution)
Tamil Nadu
India

V. VANITHA
Karpagam Academy of Higher Education
Coimbatore
India

S. VANSHIKA
Department of Computer Science and Engineering
Chandigarh University
Gharuan
India

V. VELMURUGAN
Department of EEE
Arunai Engineering College
Tiruvannamalai
India

M. YASHICA
Karpagam Academy of Higher
Education
Coimbatore
India

Index

O, P

R, S

T, V, W

Other titles from

ISTE

in

Computer Engineering

2026

BHAMBRI Pankaj, DUTTA Pushan Kumar, KHAN Mudassir, STAROSTKA-PATYK Marta
Next-Gen Supply Chains: AI, Automation, and Sustainability in a Disrupted World

DARCHE Philippe
Computer Memories 1: Core Concepts
Computer Memories 2: Static Random-Access Memory

KUMAR Abhishek, BATTA Priya, AHUJA Sachin, RATHORE Pramod Singh
AI-driven Innovations in Physiotherapy and Oncology 3

KUMAR Abhishek, BATTA Priya, ANANTH J.P.
AI-driven Healthcare Innovations: Applications in Neurology and Medicine

MACHADO Carolina, DAVIM J. Paulo
Artificial Intelligence Technologies in Management and Engineering

MAHOUT Vincent, ROCACHER Thierry, AURIOL Guillaume
Programming of Embedded Systems: Applications to SM32F10x µControllers

2025

CHIKH Azeddine
Systemic Approach to Categorizing and Modeling Requirements

KUMAR Abhishek, BATTA Priya, AHUJA Sachin,
RATHORE Pramod Singh
AI-driven Innovations in Physiotherapy and Oncology 1
AI-driven Innovations in Physiotherapy and Oncology 2

TANEJA Ashu, KUMAR Abhishek, LIMKAR Suresh Vishnudas,
OUAISSA Mariya, OUAISSA Mariyam
Driving Innovation through AI and Digital Twin for 6G Powered Sustainable Ultra Smart Cities

TANEJA Ashu, KUMAR Abhishek, LIMKAR Suresh Vishnudas,
OUAISSA Mariya, OUAISSA Mariyam
6G Urban Innovation: AI and Digital Twin for Next-Gen Sustainable Cities

2024

CHAKRABORTY Rajdeep, GHOSH Anupam, KUMAR MANDAL Jyotsna,
CHOUDHURY Tanupriya, CHATTERJEE Prasenjit
Smart Edge Computing: An Operation Research Perspective
(International Perspectives in Decision Analytics and Operations Research Set – Volume 2)

HOMÈS Bernard
Fundamentals of Software Testing (2nd edition, revised and updated)

JERBI Abdessalem
Discrete-Event Simulation: Concepts and Production in Arena

PRADHAN Devasis, KUMAR SAHU Prasanna, MYO TUN Hla,
CHATTERJEE Prasenjit
Artificial and Cognitive Computing for Sustainable Healthcare Systems in Smart Cities
(International Perspectives in Decision Analytics and Operations Research Set – Volume 3)

2023

CHOUDHURY Tanupriya, KHANNA Abhirup, CHATTERJEE Prasenjit,
UM Jung-Sup, BHATTACHARYA Abhishek
Blockchain Applications in Healthcare: Innovations and Practices (International Perspectives in Decision Analytics and Operations Research Set – Volume 1)

PAI G A Vijayalakshmi
A Textbook of Data Structures and Algorithms 1: Mastering Linear Data Structures
A Textbook of Data Structures and Algorithms 2: Mastering Nonlinear Data Structures
A Textbook of Data Structures and Algorithms 3: Mastering Advanced Data Structures and Algorithm Design Strategies

2022

HOMÈS Bernard
Advanced Testing of Systems-of-Systems 1: Theoretical Aspects
Advanced Testing of Systems-of-Systems 2: Practical Aspects

MEHTA Shikha, TIWARI Sanju, SIARRY Patrick, JABBAR M.A.
Tools, Languages, Methodologies for Representing Semantics on the Web of Things

SIDHOM Sahbi, KADDOUR Amira
Systems and Uses of Digital Sciences for Knowledge Organization (Digital Tools and Uses Set – Volume 9)

ZAIDOUN Ameur Salem
Computer Science Security: Concepts and Tools

2021

DARCHE Philippe
Microprocessor 4: Core Concepts – Software Aspects
Microprocessor 5: Software and Hardware Aspects of Development, Debugging and Testing – The Microcomputer

DELHAYE Jean-Loïc
Inside the World of Computing: Technologies, Uses, Challenges

HARDIN Thérèse, JAUME Mathieu, PESSAUX François,
VIGUIÉ DONZEAU-GOUGE Véronique
Concepts and Semantics of Programming Languages 1: A Semantical Approach with OCaml and Python
Concepts and Semantics of Programming Languages 2: Modular and Object-oriented Constructs with OCaml, Python, C++, Ada and Java

MKADMI Abderrazak
Archives in The Digital Age: Preservation and the Right to be Forgotten (Digital Tools and Uses Set – Volume 8)

TOKLU Yusuf Cengiz, BEKDAŞ Gebrail, NIGDELI Sinan Melih
Metaheuristics for Structural Design and Analysis (Optimization Heuristics Set – Volume 3)

2020

DARCHE Philippe
Microprocessor 1: Prolegomenes – Calculation and Storage Functions – Models of Computation and Computer Architecture
Microprocessor 2: Core Concepts – Communication in a Digital System
Microprocessor 3: Core Concepts – Hardware Aspects

DUVAUT Patrick, DALLOZ Xavier, MENGA David, KOEHL François,
CHRIQUI Vidal, BRILL Joerg
Internet of Augmented Me, I.AM: Empowering Innovation for a New Sustainable Future

LAFFLY Dominique
TORUS 1 – Toward an Open Resource Using Services: Cloud Computing for Environmental Data
TORUS 2 – Toward an Open Resource Using Services: Cloud Computing for Environmental Data
TORUS 3 – Toward an Open Resource Using Services: Cloud Computing for Environmental Data

LAURENT Anne, LAURENT Dominique, MADERA Cédrine
Data Lakes
(Databases and Big Data Set – Volume 2)

OULHADJ Hamouche, DAACHI Boubaker, MENASRI Riad
Metaheuristics for Robotics
(Optimization Heuristics Set – Volume 2)

SADIQUI Ali
Computer Network Security

VENTRE Daniel
Artificial Intelligence, Cybersecurity and Cyber Defense

2019

BESBES Walid, DHOUIB Diala, WASSAN Niaz, MARREKCHI Emna
Solving Transport Problems: Towards Green Logistics

CLERC Maurice
Iterative Optimizers: Difficulty Measures and Benchmarks

GHLALA Riadh
Analytic SQL in SQL Server 2014/2016

TOUNSI Wiem
Cyber-Vigilance and Digital Trust: Cyber Security in the Era of Cloud Computing and IoT

2018

ANDRO Mathieu
Digital Libraries and Crowdsourcing
(Digital Tools and Uses Set – Volume 5)

ARNALDI Bruno, GUITTON Pascal, MOREAU Guillaume
Virtual Reality and Augmented Reality: Myths and Realities

CARDON Alain
Beyond Artificial Intelligence: From Human Consciousness to Artificial Consciousness

HOMAYOUNI S. Mahdi, FONTES Dalila B.M.M.
Metaheuristics for Maritime Operations
(Optimization Heuristics Set – Volume 1)

JEANSOULIN Robert
JavaScript and Open Data

PIVERT Olivier
NoSQL Data Models: Trends and Challenges
(Databases and Big Data Set – Volume 1)

SEDKAOUI Soraya
Data Analytics and Big Data

SALEH Imad, AMMI Mehdi, SZONIECKY Samuel
Challenges of the Internet of Things: Technology, Use, Ethics
(Digital Tools and Uses Set – Volume 7)

SZONIECKY Samuel
Ecosystems Knowledge: Modeling and Analysis Method for Information and Communication
(Digital Tools and Uses Set – Volume 6)

2017

BENMAMMAR Badr
Concurrent, Real-Time and Distributed Programming in Java: Threads, RTSJ and RMI

BOUHAÏ Nasreddine, SALEH Imad
Internet of Things: Evolutions and Innovations
(Digital Tools and Uses Set – Volume 4)

HÉLIODORE Frédéric, NAKIB Amir, ISMAIL Boussaad, OUCHRAA Salma, SCHMITT Laurent
Metaheuristics for Intelligent Electrical Networks
(Metaheuristics Set – Volume 10)

MA Haiping, SIMON Dan
Evolutionary Computation with Biogeography-based Optimization
(Metaheuristics Set – Volume 8)

PÉTROWSKI Alain, BEN-HAMIDA Sana
Evolutionary Algorithms
(Metaheuristics Set – Volume 9)

PAI G A Vijayalakshmi
Metaheuristics for Portfolio Optimization: An Introduction using MATLAB®
(Metaheuristics Set – Volume 11)

REYES-GARCIA Everardo
The Image-Interface: Graphical Supports for Visual Information
(Digital Tools and Uses Set – Volume 3)

REYES-GARCIA Everardo, BOUHAÏ Nasreddine
Designing Interactive Hypermedia Systems
(Digital Tools and Uses Set – Volume 2)

SZONIECKY Samuel, BOUHAÏ Nasreddine
Collective Intelligence and Digital Archives: Towards Knowledge Ecosystems
(Digital Tools and Uses Set – Volume 1)

2016

BLUM Christian, FESTA Paola
Metaheuristics for String Problems in Bio-informatics
(Metaheuristics Set – Volume 6)

DEROUSSI Laurent
Metaheuristics for Logistics
(Metaheuristics Set – Volume 4)

DHAENENS Clarisse, JOURDAN Laetitia
Metaheuristics for Big Data
(Metaheuristics Set – Volume 5)

LABADIE Nacima, PRINS Christian, PRODHON Caroline
Metaheuristics for Vehicle Routing Problems
(Metaheuristics Set – Volume 3)

LEROY Laure
Eyestrain Reduction in Stereoscopy

LUTTON Evelyne, PERROT Nathalie, TONDA Alberto
Evolutionary Algorithms for Food Science and Technology
(Metaheuristics Set – Volume 7)

MAGOULÈS Frédéric, ZHAO Hai-Xiang
Data Mining and Machine Learning in Building Energy Analysis

RIGO Michel
Advanced Graph Theory and Combinatorics

ZELM Martin, DOUMEINGTS Guy, MENDONÇA Joao Pedro
Enterprise Interoperability: in the Digitized and Networked Factory of the Future

2015

BARBIER Franck, RECOUSSINE Jean-Luc
COBOL Software Modernization: From Principles to Implementation with the BLU AGE® Method

CHEN Ken
Performance Evaluation by Simulation and Analysis with Applications to Computer Networks

CLERC Maurice
Guided Randomness in Optimization
(Metaheuristics Set – Volume 1)

DURAND Nicolas, GIANAZZA David, GOTTELAND Jean-Baptiste, ALLIOT Jean-Marc
Metaheuristics for Air Traffic Management
(Metaheuristics Set – Volume 2)

MAGOULÈS Frédéric, ROUX François-Xavier, HOUZEAUX Guillaume
Parallel Scientific Computing

MUNEESAWANG Paisarn, YAMMEN Suchart
Visual Inspection Technology in the Hard Disk Drive Industry

2014

BOULANGER Jean-Louis
Formal Methods Applied to Industrial Complex Systems
Formal Methods Applied to Complex Systems: Implementation of the B Method

EXPOSITO Ernesto, DIOP Codé
Smart SOA Platforms in Cloud Computing Architectures

GARDI Frédéric, BENOIST Thierry, DARLAY Julien, ESTELLON Bertrand, MEGEL Romain
Mathematical Programming Solver based on Local Search

KRICHEN Saoussen, CHAOUACHI Jouhaina
Graph-related Optimization and Decision Support Systems

LARRIEU Nicolas, VARET Antoine
Rapid Prototyping of Software for Avionics Systems: Model-oriented Approaches for Complex Systems Certification

LAURAS Matthieu, ZELM Martin, ARCHIMÈDE Bernard, BÉNABEN Frédérick, DOUMEINGTS Guy
Enterprise Interoperability: Interoperability for Agility, Resilience and Plasticity of Collaborations

OUSSALAH Mourad Chabane
Software Architecture 1
Software Architecture 2

QUESNEL Flavien
Scheduling of Large-scale Virtualized Infrastructures: Toward Cooperative Management

RIGO Michel
Formal Languages, Automata and Numeration Systems 1: Introduction to Combinatorics on Words
Formal Languages, Automata and Numeration Systems 2: Applications to Recognizability and Decidability

SAINT-DIZIER Patrick
Musical Rhetoric: Foundations and Annotation Schemes

TOUATI Sid, DUPONT DE DINECHIN Benoit
Advanced Backend Code Optimization

2013

ANDRÉ Étienne, SOULAT Romain
The Inverse Method: Parametric Verification of Real-time Embedded Systems

BOULANGER Jean-Louis
Safety Management for Software-based Equipment

DELAHAYE Daniel, PUECHMOREL Stéphane
Modeling and Optimization of Air Traffic

FRANCOPOULO Gil
LMF — Lexical Markup Framework

GHEDIRA Khaled
Constraint Satisfaction Problems: CSP Formalisms and Techniques

GOVAERT Gérard, NADIF Mohamed
Co-Clustering: Models, Algorithms and Applications

LANDRAGIN Frédérick
Man–Machine Dialogue: Design and Challenges

LUCAS Laurent, LOSCOS Céline, REMION Yannick
3D Video: Models, From Capture to Diffusion

ROCHANGE Christine, UHRIG Sascha, SAINRAT Pascal
Time-Predictable Architectures

WAHBI Mohamed
Algorithms and Ordering Heuristics for Distributed Constraint Satisfaction Problems

ZELM Martin, VAN SINDEREN Marten, FERRAIRA PIRES Luis, DOUMEINGTS Guy
Enterprise Interoperability: Research and Applications in the Service-oriented Ecosystem

2012

ARBOLEDA Hugo, ROYER Jean-Claude
Model-Driven and Software Product Line Engineering

BLANCHET Gérard, DUPOUY Bertrand
Computer Architecture

BOULANGER Jean-Louis
Industrial Use of Formal Methods: Formal Verification
Formal Methods: Industrial Use from Model to the Code

CALVARY Gaëlle, DELOT Thierry, SÈDES Florence, TIGLI Jean-Yves
Computer Science and Ambient Intelligence

MAHOUT Vincent
Assembly Language Programming: ARM Cortex-M3 2.0: Organization, Innovation and Territory

MARLET Renaud
Program Specialization

SOTO Maria, SEVAUX Marc, ROSSI André, LAURENT Johann
Memory Allocation Problems in Embedded Systems: Optimization Methods

2011

BICHOT Charles-Edmond, SIARRY Patrick
Graph Partitioning

BOULANGER Jean-Louis
Static Analysis of Software: The Abstract Interpretation

CAFERRA Ricardo
Logic for Computer Science and Artificial Intelligence

KORDON Fabrice, HADDAD Serge, PAUTET Laurent, PETRUCCI Laure
Distributed Systems: Design and Algorithms
Models and Analysis in Distributed Systems

LORCA Xavier
Tree-based Graph Partitioning Constraint

RIDHA MAHJOUB A.
Progress in Combinatorial Optimization: ISCO2010

TRUCHET Charlotte, ASSAYAG Gerard
Constraint Programming in Music

VICAT-BLANC Pascale, GOGLIN Brice, GUILLIER Romaric, SOUDAN Sébastien
Computing Networks: From Cluster to Cloud Computing

2010

AUBRUN Christophe, SIMON Daniel, SONG Ye-Qiong
Co-design Approaches for Dependable Networked Control Systems

AUDIBERT Pierre
Mathematics for Informatics and Computer Science

BABAU Jean-Philippe, BLAY-FORNARINO Mireille, CHAMPEAU Joël, ROBERT Sylvain, SABETTA Antonino
Model Driven Engineering for Distributed Real-Time Systems: MARTE Modeling, Model Transformations and their Usages

BOULANGER Jean-Louis
Safety of Computer Architectures

SIGAUD Olivier, BUFFET Olivier
Markov Decision Processes in Artificial Intelligence

MONMARCHÉ Nicolas, GUINAND Frédérick, SIARRY Patrick
Artificial Ants

SOLNON Christine
Ant Colony Optimization and Constraint Programming

2009

FOURNIER Jean-Claude
Graph Theory and Applications: with exercises and problems

GUÉDON Jeanpierre
The Mojette Transform: theory and applications

JARD Claude, ROUX Olivier H.
Communicating Embedded Systems: Software and Design

LECOUTRE Christophe
Constraint Networks: Techniques and Algorithms

2008

BANÂTRE Michel, MARRON Pedro Jose, OLLERO Anibal, WOLITZ Adam
Cooperating Embedded Systems and Wireless Sensor Networks

MERZ Stephan, NAVET Nicolas
Modeling and Verification of Real-time Systems: Formalisms and Software Tools

PASCHOS Vangelis Th.
Combinatorial Optimization and Theoretical Computer Science: Interfaces and Perspectives

WALDNER Jean-Baptiste
Nanocomputers and Swarm Intelligence

2007

BENHAMOU Frédéric, JUSSIEN Narendra, O'SULLIVAN Barry
Trends in Constraint Programming

JUSSIEN Narendra
A to Z of Sudoku

2006

BABAU Jean-Philippe, CHAMPEAU Joël, GÉRARD Sébastien
From MDD Concepts to Experiments and Illustrations – DRES 2006

HABRIAS Henri, FRAPPIER Marc
Software Specification Methods: An Overview Using a Case Study

MURAT Cécile, PASCHOS Vangelis Th.
Probabilistic Combinatorial Optimization on Graphs

PANETTO Hervé, BOUDJLIDA Nacer
Interoperability for Enterprise Software and Applications: Proceedings of the Workshops and the Doctorial Symposium of the Second IFAC/IFIP I-ESA International Conference: EI2N, WSI, IS-TSPQ 2006

2005

GÉRARD Sébastien, BABAU Jean-Philippe, CHAMPEAU Joël
Model Driven Engineering for Distributed Real Time Embedded Systems

PANETTO Hervé
Interoperability of Enterprise Software and Applications: Workshops of the INTEROP-ESA 2005 International Conference (EI2N, WSI, ISIDI and IEHENA 2005)

Printed and bound by CPI Group (UK) Ltd, Croydon, CR0 4YY
25/06/2026
14907692-0004